乳清蛋白质化学与应用
Whey Protein Production, Chemistry, Functionality and Applications

〔美〕郭明若 著

郭明若 王喜波 等 编译

科学出版社

北京

图字：01-2021-3451 号

内 容 简 介

乳清蛋白质具有营养丰富、功能特性优良的特征，乳清蛋白质产品是乳品工业最重要的原料之一，并广泛用于制药、消费品、生物材料、环保制品等领域。本书主要内容包括乳清蛋白质生产及其发展、生产工艺技术、化学特性、结构和变性、与其他食品组分的相互作用、营养特性、改性、应用及展望等。本书为我国乳清的综合开发和高附加值利用提供了理论和技术参考。

本书将英文原版编译成中文版，供乳品领域专家学者、高校乳品专业师生、企业研发技术及管理人员阅读参考。

All Right Reserved. Authorised Translation from the English Language edition published by John Wiley & Sons Limited. Responsibility for the accuracy of the translation rests solely with China Science Publishing & Media Ltd.（Science Press） and is not the responsibility of John Wiley & Sons Limited. No part of this book may be reproduced in any form without the written permission of the original copyright holder, John Wiley & Sons Limited.

图书在版编目（CIP）数据

乳清蛋白质化学与应用 /（美）郭明若（Mingruo Guo）著；郭明若等编译. —北京：科学出版社，2022.3
书名原文：Whey Protein Production, Chemistry, Functionality and Applications
原书 ISBN 978-1-119-25602-1
ISBN 978-7-03-071506-7

Ⅰ. ①乳… Ⅱ. ①郭… Ⅲ. ①乳蛋白–应用化学 Ⅳ. ①Q513

中国版本图书馆 CIP 数据核字（2022）第 028204 号

责任编辑：贾 超 孙静惠 / 责任校对：杜子昂
责任印制：吴兆东 / 封面设计：东方人华

科学出版社 出版
北京东黄城根北街 16 号
邮政编码：100717
http://www.sciencep.com

北京中石油彩色印刷有限责任公司 印刷
科学出版社发行 各地新华书店经销

*

2022 年 3 月第 一 版　开本：720×1000　1/16
2022 年 3 月第一次印刷　印张：13 3/4
字数：270 000

定价：128.00 元
（如有印装质量问题，我社负责调换）

本书编委会

主　编：郭明若　王喜波

副主编：姜云庆　孙晓萌　孙玉雪

编者（排名不分先后）：

郭明若　程建军　刘迪茹　王　睦

王　旭　王翠娜　王国荣　姜云庆

孙玉雪　孙晓萌　王喜波

前　言

干酪是复杂且古老的食品,其发展历史可以追溯到几千年前。它也是一种主要的乳制品,在人类的营养和乳品发展方面起到重要作用。干酪生产是从牛乳中分离出酪蛋白和脂肪并浓缩加工的过程,乳清是干酪制品的副产物。随着乳品加工新技术的发展,几乎所有的乳清成分(特别是乳清蛋白质)都得到了充分的开发利用。

乳清蛋白质营养价值高、功能性好,是乳品工业最重要的原料之一,不仅应用于食品,还应用于制药、消费品、生物材料、环保制品等。由于干酪产业在中国刚刚起步,乳清产品尤其是乳清蛋白质产品几乎全部依赖进口,而乳清蛋白质又是生产婴儿配方奶粉等产品的关键原料之一,因此,构建完善的乳品工业体系,改善产品结构,保障乳清蛋白质产品生产与供给,具有一定的战略意义。期望本书中文版的出版,对我国乳清蛋白质等相关领域的科研、生产和教学等方面能够起到抛砖引玉之效。

我从事食品化学和乳品开发领域研究三十多年,其中二十余年的研究集中在乳清的综合利用和乳清蛋白质的功能性质与应用方面。本书是我所在实验室在乳清加工科学与技术研究方面取得成果的总结(包括研究成果、出版物、专利和未公开发表的数据等),重点论述乳清蛋白质的生产、化学特性、功能特性和实践应用等。全书共10章,包括乳清蛋白质的生产及其发展、生产工艺技术、化学特性、结构和变性以及与其他食品组分的相互作用、营养特性、在营养产品中的应用、功能特性及其在食品中的应用、改性、在非食品领域的应用、展望等。

衷心感谢参与本书撰写和编译的同仁们,他们的辛勤努力与付出使本书顺利出版,特别是东北农业大学王喜波教授在中文版编写过程中作为联系人与出版社沟通,在译稿、统稿、校对等方面做了大量工作;同时也要感谢石佳卉、李宁、张智慧、孙小童、宋天睿、刘恒霖、于海洋、郝良焕等研究生在文字和排版方面提供的帮助。

我还要感谢科学出版社贾超先生在本书出版过程中的帮助,感谢John Wiley & Sons出版公司生命科学与地球科学部的David McDade先生在本书英文版发表过程中的帮助,感谢Athira Menon女士对本书英文版编辑方面的协助。

最后，我要感谢家人（尤其谢谢我的妻子 Ying），感谢他们对我的爱和对我繁忙的学术工作、早起晚归的工作习惯（即使是在假期）的理解。

美国伯灵顿

2022 年 3 月

目 录

前言
第1章 乳清蛋白质生产及其发展 ··· 1
  1.1 乳清蛋白质种类 ··· 1
    1.1.1 甜乳清 ·· 2
    1.1.2 酸乳清 ·· 3
  1.2 乳清的利用 ··· 5
    1.2.1 古代时期 ·· 5
    1.2.2 工业革命时期 ·· 5
    1.2.3 现代发展 ·· 6
  1.3 乳清产品概况 ··· 6
    1.3.1 乳糖 ·· 6
    1.3.2 乳清粉 ·· 6
    1.3.3 乳清浓缩蛋白和乳清分离蛋白 ································· 6
    1.3.4 其他乳清蛋白质产品 ·· 7
    1.3.5 乳矿物质 ·· 7
  1.4 总结 ··· 7
  参考文献 ·· 7

第2章 乳清蛋白质产品的生产工艺技术 ····································· 11
  2.1 乳清蛋白质回收技术 ··· 11
    2.1.1 热/酸沉淀 ··· 11
    2.1.2 膜加工技术 ·· 12
  2.2 乳清蛋白质组分的分离 ··· 15
    2.2.1 α-乳白蛋白和β-乳球蛋白的分离 ······························· 15
    2.2.2 GMP 分离 ·· 19
    2.2.3 牛血清白蛋白和免疫球蛋白分离 ······························· 20
    2.2.4 乳铁蛋白和乳过氧化物酶分离 ································· 20
  2.3 乳清产品加工 ··· 22
    2.3.1 液体乳清的澄清、分离和巴氏杀菌 ····························· 22
    2.3.2 膜过滤 ·· 23

        2.3.3　脱盐 ·· 24
        2.3.4　浓缩 ·· 24
        2.3.5　干燥 ·· 24
    2.4　总结 ·· 25
    参考文献 ·· 25
第3章　乳清蛋白质的化学特性 ·· 31
    3.1　β-乳球蛋白 ·· 31
        3.1.1　β-乳球蛋白的化学特性 ··· 31
        3.1.2　β-乳球蛋白的分离与制备 ··· 34
        3.1.3　β-乳球蛋白的生物特性 ··· 35
    3.2　α-乳白蛋白 ·· 37
        3.2.1　α-乳白蛋白的化学特性 ··· 37
        3.2.2　α-乳白蛋白的分离 ·· 39
        3.2.3　α-乳白蛋白的功能特性 ··· 40
    3.3　牛血清白蛋白 ·· 41
    3.4　乳铁蛋白 ·· 42
    3.5　免疫球蛋白 ·· 44
    3.6　微量蛋白质 ·· 45
        3.6.1　生长因子 ·· 45
        3.6.2　乳过氧化物酶 ·· 45
        3.6.3　乳脂肪球膜蛋白 ·· 46
        3.6.4　维生素结合蛋白 ·· 47
    3.7　总结 ·· 47
    参考文献 ·· 47
第4章　乳清蛋白质的结构和变性以及与其他食品组分的相互作用 ··································· 55
    4.1　乳清蛋白质的结构和变性 ·· 55
        4.1.1　热变性 ·· 55
        4.1.2　乳清蛋白质的酶促变性 ·· 57
        4.1.3　乳清蛋白质的超声改性 ·· 57
        4.1.4　乳清蛋白质的辐射改性 ·· 58
    4.2　巯基和二硫键在乳清蛋白质聚集和凝胶中的作用 ·· 59
        4.2.1　巯基和二硫键在乳清蛋白质聚集中的作用 ·· 59
        4.2.2　巯基和二硫键在乳清蛋白质凝胶中的作用 ·· 60
    4.3　乳清蛋白质和酪蛋白的相互作用 ·· 61
        4.3.1　模型系统中乳清蛋白质和酪蛋白的相互作用 ·· 61

  4.3.2　乳清蛋白质和酪蛋白微粒的相互作用 …………………………… 61
 4.4　乳清蛋白质和糖类的相互作用 …………………………………………… 62
  4.4.1　乳清蛋白质和糖类之间的美拉德反应 …………………………… 62
  4.4.2　乳清蛋白质和多糖的相互作用 …………………………………… 64
 4.5　乳清蛋白质和食品中其他成分的相互作用 ……………………………… 70
  4.5.1　明胶 …………………………………………………………………… 70
  4.5.2　卵磷脂 ………………………………………………………………… 71
 4.6　总结 ………………………………………………………………………… 72
 参考文献 …………………………………………………………………………… 72

第 5 章　乳清蛋白质的营养特性 ……………………………………………………… 83
 5.1　氨基酸组成：乳清蛋白质与母乳蛋白 …………………………………… 83
 5.2　乳清蛋白质中的支链氨基酸 ……………………………………………… 84
 5.3　乳清蛋白质衍生物 ………………………………………………………… 85
 5.4　乳清蛋白质的致敏性和消化率 …………………………………………… 85
 5.5　乳清蛋白质组分的治疗特性 ……………………………………………… 86
  5.5.1　糖尿病 ………………………………………………………………… 86
  5.5.2　癌症 …………………………………………………………………… 87
  5.5.3　肝病 …………………………………………………………………… 88
  5.5.4　心血管疾病 …………………………………………………………… 89
  5.5.5　免疫系统疾病 ………………………………………………………… 89
 5.6　乳清蛋白质的抗氧化特性 ………………………………………………… 90
  5.6.1　乳清蛋白质的抗氧化活性 …………………………………………… 90
  5.6.2　乳清蛋白质组分的抗氧化活性 ……………………………………… 91
  5.6.3　乳清蛋白质抗氧化肽 ………………………………………………… 95
  5.6.4　乳清蛋白质抗氧化活性在食品中的应用 …………………………… 97
 5.7　总结 ………………………………………………………………………… 100
 参考文献 …………………………………………………………………………… 101

第 6 章　乳清蛋白质在营养产品中的应用 …………………………………………… 112
 6.1　婴幼儿配方奶粉 …………………………………………………………… 113
  6.1.1　乳清蛋白质与酪蛋白比例 …………………………………………… 113
  6.1.2　婴幼儿配方奶粉的配方及加工 ……………………………………… 114
  6.1.3　新一代婴幼儿配方奶粉的乳清蛋白质 ……………………………… 115
 6.2　运动营养品 ………………………………………………………………… 116
  6.2.1　蛋白质代谢 …………………………………………………………… 116
  6.2.2　罐装乳清蛋白质 ……………………………………………………… 117

  6.2.3 酸化乳清蛋白质营养饮料 ································· 118
  6.2.4 蛋白质能量补充产品 ··································· 119
6.3 老年人蛋白质补充剂 ·· 120
6.4 代餐食品 ·· 121
6.5 高蛋白共生酸奶 ·· 122
6.6 总结 ·· 123
参考文献 ·· 123

第7章 乳清蛋白质功能特性及其在食品中的应用 ··············· 127
7.1 作为食品增稠剂/胶凝剂 ·· 127
7.2 作为食品稳定剂/乳化剂 ·· 129
  7.2.1 乳清蛋白质乳液的表征技术 ······························· 129
  7.2.2 乳清蛋白质乳液的制备 ··································· 130
  7.2.3 乳清蛋白质乳液的稳定性 ································· 132
  7.2.4 乳清蛋白质/水胶体乳液稳定性 ···························· 133
  7.2.5 其他乳化剂对乳清蛋白质乳液稳定性的影响 ················· 134
7.3 脂肪或乳品替代品 ·· 134
7.4 疏水性营养食品载体 ·· 135
  7.4.1 类胡萝卜素 ·· 135
  7.4.2 多酚 ·· 136
7.5 微囊化壁材 ·· 137
  7.5.1 乳清蛋白质基风味化合物和脂质微胶囊的制备 ··············· 138
  7.5.2 益生菌微囊化 ·· 141
  7.5.3 微囊化益生菌在食品中的应用 ····························· 143
  7.5.4 生物活性成分的微囊化 ··································· 144
7.6 食用膜和涂层 ·· 145
  7.6.1 食用膜和涂层参数 ······································ 145
  7.6.2 乳清蛋白质薄膜/涂层 ··································· 146
  7.6.3 乳清蛋白质可食用膜/涂层的组成 ·························· 147
  7.6.4 乳清蛋白质/多糖复合膜的物理性质 ························ 150
  7.6.5 乳清蛋白质膜在食品工业中的应用 ························· 150
7.7 总结 ·· 153
参考文献 ·· 153

第8章 乳清蛋白质的改性 ··································· 164
8.1 热处理 ·· 164
8.2 酶处理 ·· 166

	8.2.1 转谷氨酰胺酶交联	166
	8.2.2 酶水解	168
8.3	超声波处理	168
8.4	高压处理	169
8.5	电脉冲处理	171
8.6	辐射处理	172
	8.6.1 γ射线辐照	172
	8.6.2 紫外线照射	172
8.7	化学改性	173
8.8	总结	174
参考文献		174

第 9 章 乳清蛋白质在非食品领域的应用 ... 182

9.1	胶水黏合理论	182
	9.1.1 吸附理论	183
	9.1.2 机械互锁理论	183
	9.1.3 化学键合理论	183
9.2	木清漆	184
9.3	木胶水	185
9.4	办公胶水	192
9.5	手术胶	195
9.6	总结	196
参考文献		196

第 10 章 乳清蛋白质展望 ... 199

10.1	乳清蛋白质市场需求	199
10.2	希腊酸奶的盛行和酸乳清	199
10.3	微滤乳和清蛋白	201
10.4	总结	203
参考文献		203

索引 ... 206

第 1 章　乳清蛋白质生产及其发展

Mingruo Guo[1,2] and Guorong Wang[1]

1. Department of Nutrition and Food Science, University of Vermont, Burlington, USA
2. College of Food Science, Northeast Agriculture University, Harbin, People's Republic of China

1.1　乳清蛋白质种类

牛奶是由脂肪球、酪蛋白微粒和清蛋白或乳清相组成的复杂的胶体悬浮溶液（图 1.1）。乳清是牛奶由凝乳酶或酸凝固后压滤获得的一种黄色到绿色的透明溶液。乳清组分是指那些不参与凝乳并能被滤出的小分子成分。乳清固形物包括乳糖、蛋白质（主要是乳清蛋白质）和矿物质等，如表 1.1 所示。液体乳清固形物占全乳固形物的 50%左右，包括大多数矿物质、几乎全部的乳清蛋白质和乳糖。

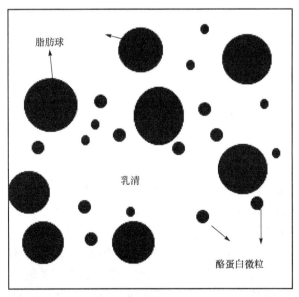

图 1.1　牛奶是由脂肪球、酪蛋白微粒和清蛋白或乳清相组成的一个复杂的胶体悬浮溶液。乳清蛋白质、乳糖和矿物质存在于乳清相中

表 1.1　乳和乳清的组成

	全乳	乳清
酪蛋白（%，w/v）	2.8	<0.1
乳清蛋白质（%，w/v）	0.7	0.7
脂肪（%，w/v）	3.7	0.1
灰分（%，w/v）	0.7	0.5
乳糖（%，w/v）	4.9	4.9
总固形物（%，w/v）	12.8	6.3

注：资料来自 Smithers（2008）；w/v 表示质量浓度。

不同方法凝固的牛奶会产生不同类型的乳清。一般可分为甜乳清和酸乳清。甜乳清和酸乳清没有明确的定义来区分，但通常以 pH 5.6 为界，甜乳清 pH 高于 5.6，酸乳清 pH 低于 5.6。甜乳清来源于干酪生产（凝乳酶凝乳工艺），有时也称为干酪乳清；酸乳清来源于发酵凝乳工艺（乳糖转化为乳酸，如希腊酸奶）或酸凝乳工艺（酸法生产酪蛋白）（Tunick，2008）。甜乳清和酸乳清的组成列于表 1.2。

表 1.2　甜乳清和酸乳清的组成

	甜乳清	酸乳清
蛋白质（g/L）	6～10	6～8
乳糖（g/L）	46～52	44～46
矿物质（g/L）	2.5～4.7	4.3～7.2
pH	>5.6	<5.6

注：资料来自 Tunick（2008）。

1.1.1　甜乳清

牛、绵羊和山羊等哺乳动物已被驯化超过 10000 年（Clutton-Brock，1999；Beja-Pereira et al.，2006）。DNA 分析技术表明哺乳动物被驯化历史可以追溯到 17000 年以前（Beja-Pereira et al.，2006），驯化的牛和其他哺乳动物可以用来耕地以及提供奶类、皮毛和肉，但牛在东亚、中非地区却没有用来提供牛奶（Clutton-Brock，1999）。直到现在，这些地区患乳糖不耐受的人数远比北欧和近东等地区的多。史前时期，在不屠宰珍贵牲畜的情况下，奶的生产非常重要，奶可持续提供营养食物。干酪的制造是人类文明史上的里程碑，它比新鲜牛奶更容易保存。古今干酪制造工艺有许多共同点，包括自然发酵、熬煮、挤压和干燥。

第一块干酪可能来自反刍动物的胃（Smithers，2008），胃中天然存在的凝乳酶使牛奶凝成凝乳，然后凝乳经挤压分离乳清，这可能是人类第一次获得乳清。世界各地的考古工作者发现了古时利用牛奶的证据（陶器中牛奶/干酪残留物）（Evershed et al.，2008；Salque et al.，2013；Scott et al.，1998；Yang et al.，2014）。北欧最早制作干酪的证据是公元前 6000 年（Salque et al.，2013）的压榨乳清的陶瓷筛子碎片[图 1.2（a）]、修复的筛子容器碎片[图 1.2（b）]与法国上卢瓦尔的现代干酪筛非常相似（Briggs，2012）。

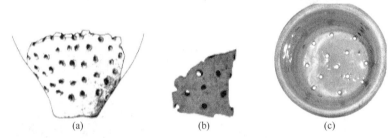

图 1.2　（a）复原的筛子、（b）发现于波兰 Kuyavia 地区的筛子碎片（7000 年历史）、（c）现代用于干酪压滤的陶瓷滤锅

典型的干酪生产包括添加凝乳酶以破坏酪蛋白微粒丝（κ-酪蛋白丝），进而破坏蛋白胶束结构，促进牛奶凝固（O'Callaghan et al.，2002），再通过切割和挤压凝乳块滤出乳清。凝乳酶凝乳过程如图 1.3 所示。凝乳酶是来源于反刍动物胃中的复合酶，它可以切断κ-酪蛋白丝（一种稳定酪蛋白微粒结构的蛋白），从而使酪蛋白微粒凝固（Daviau et al.，2000）。脂肪球被酪蛋白凝乳包裹或乳化；而乳清则挤压滤出，称为甜乳清或干酪乳清。生产 1 份干酪同时产生 9 份液体乳清。由凝乳酶切割的κ-酪蛋白片段称为糖巨肽（GMP）（Brody，2000），通常存在于甜乳清中。凝乳酶不能将乳糖转化为乳酸，因而得到中性 pH 的甜乳清。目前全球干酪产量巨大，其副产物甜乳清成为主要的商品乳清。

1.1.2　酸乳清

酸乳清是生产酸性酪蛋白和希腊酸奶等酸凝乳产品的副产物。中性 pH 时，κ-酪蛋白丝（通过静电斥力）和磷酸钙胶体（CCP）使酪蛋白微粒状态稳定（de Kruif and Holt，2003）。酸凝乳机制如图 1.4 所示。当 pH 下降时，一方面，κ-酪蛋白静电斥力被中和，并导致胶束收缩（de Kruif，1997）；另一方面，与酪蛋白分子结合的磷酸钙胶体在乳清相中溶解（Le Graët and Gaucheron，1999），酪蛋白微粒被破坏形成凝乳（Lucey，2003），从酸乳中压滤出的乳清称为酸乳清。酸凝乳过程可以通过添加无机酸或有机酸（如盐酸或乳酸）和/或发酵作用（乳糖转化为乳酸）实现。

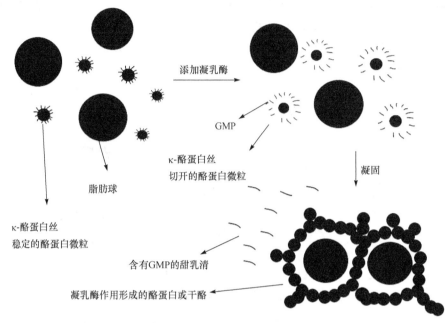

图 1.3 凝乳酶作用下的凝乳和甜乳清

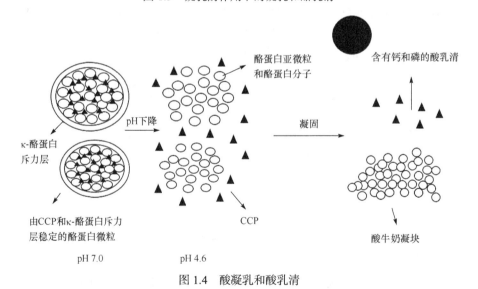

图 1.4 酸凝乳和酸乳清

凝乳酶和酸诱导的凝乳机理不同，产生的甜乳清和酸乳清的物理化学性质也有差异，如酸乳清的 pH 更低，通常不含 GMP，灰分（从胶束释放到乳清中的钙）

高，乳糖略低（一些乳糖转化为乳酸）；发酵前的热处理使酸乳清中一些乳清蛋白质（尤其是β-乳球蛋白）与κ-酪蛋白通过二硫键-巯基转换作用（Lucey，2002；Lu et al.，2013）成为凝乳的一部分，导致酸乳清中蛋白质含量降低。

1.2 乳清的利用

乳清曾经是乳制品工业中污染最严重的废弃物，其生化需氧量（BOD）是 35~45 kg/m³，化学需氧量（COD）是 60~70 kg/m³（Mawson，1994），大多数国家和地区禁止未经处理的乳清直接排放。实际上，乳清营养丰富，含有约 50%的乳固形物。乳清的开发利用进程就是变废为宝的生动写照（Smithers，2008）。

1.2.1 古代时期

铜器时代，可能从来没有考虑过如何处理乳清的问题，人们喝乳清的习惯一直延续到二十世纪初。现在，乳清用来制造各种功能性饮料，包括能量饮料（Singh and Singh，2012）、发酵饮料（Pescuma et al.，2010）、酒精饮料（Dragone et al.，2009）和碳酸饮料（Singh and Singh，2012）。早在公元前 460 年，希腊医生 Hippocrates 就给患者开了乳清药方，用于提高免疫力、治疗胃肠道疾病和皮肤病（Heffernan，2015；Smithers，2015；Susli，1956）；Smithers（2008）在文献中提到，十七世纪时乳清就用来治疗脓毒症、胃病和促进伤口愈合，那时欧洲的乳清消费成为时尚（Holsinger，1978）。从十七世纪到十九世纪，欧洲"乳清餐馆"（Whey House）菜单上的乳清粥、乳清汤、乳清茶和乳清黄油非常受欢迎（Smithers，2015）。现在，瑞士和阿尔卑斯山区豪华的乳清水疗浴场吸引着成千上万的游客。

1.2.2 工业革命时期

二十世纪初，干酪和酪蛋白生产迅速增长，导致乳清产量爆发式增长，已经超出了利用极限，只能直接排放，造成严重的环境污染，因此，学术界和工业界共同努力聚焦乳清的综合利用。

液体乳清的浓缩或干燥可以使其更容易保存或运输。1908 年，Merrell 通过喷雾干燥获得了甜乳清粉（Merrell，1911），在这个重要发明中，液体乳清通过雾化器分散为小颗粒，然后在干燥室中由"吸湿空气"干燥。早期的乳清干燥方式包括滚筒干燥（Golding and Rowsell，1932）、喷雾干燥（Merrell，1911；Peebles and Manning，1939）以及两种方式相结合（Tunick，2008）。这些干燥技术存在的问题是乳糖的吸湿性导致的能耗和成本过高，乳清蛋白质变性严重，导致溶解性和功能性差（Tunick，2008）。为了提高蛋白质溶解性，开发了多段真空蒸发器（Webb and Whittier，1948；Francis，1969），但由于缺乏从液体中回收和纯化乳清蛋白质的技术,早期的工业乳清仅用于生产动物饲料和加工乳糖(如用于婴儿食品、焙烤食品或糖果)(Berry,1923)。

1.2.3 现代发展

膜过滤分离组分是根据颗粒的大小(Zydney,1998),其机制将在第 2 章中讨论。采用膜过滤,可使乳清粉中蛋白质含量从 11%(甜乳清粉)提高到 90%[乳清分离蛋白(WPI)]。膜技术是非热加工过程,能使蛋白质热变性程度最小,液体乳清蛋白质可通过微滤或超滤回收(Morr and Ha,1993)。为了克服蒸发工艺生产浓缩乳清粉的缺陷(如溶解性差、褐变等),现代工艺采用纳滤技术浓缩原料再进行喷雾干燥生产乳清浓缩蛋白(WPC)和乳清分离蛋白(Atra et al.,2005)。

1.3 乳清产品概况

1.3.1 乳糖

乳糖是乳清中含量最高的组分,可以通过结晶方法获得,广泛应用于婴儿配方奶粉、糖果、烘焙和医药产品等(Holsinger,1988)。

1.3.2 乳清粉

液体甜乳清和酸乳清经过巴氏杀菌和喷雾干燥得到甜乳清粉和酸乳清粉(含有液体乳清所有成分),可作为乳固体替代品。乳清粉在烘焙和糖果生产中发生褐变从而产生良好色泽(Dattatreya et al.,2007),但其蛋白质含量低、灰分含量高,应用受到一定限制。利用离子交换、膜过滤或电渗析技术可去除乳清中部分盐(Houldsworth,1980)生产脱盐乳清,一般脱盐率分别为 25%、50% 和 90%,其可用于婴儿配方奶粉、酸奶和其他方面(Jost et al.,1999;Penna et al.,1997;Tratnik and Krsev,1987)。

1.3.3 乳清浓缩蛋白和乳清分离蛋白

乳清蛋白质营养丰富和功能性好,是乳清中最有价值的成分(de Wit,1998;Marshall,2004)。超滤技术可浓缩乳清蛋白质,使之占总固形物的 80%,最常见的 WPC 包括 WPC34、WPC60 和 WPC80,分别含有 34%、60% 和 80% 的蛋白质(表 1.3)。在蛋白质、乳糖和脂肪含量方面,WPC34 与脱脂奶粉组成相似,常用于替代脱脂奶粉(Hoppe et al.,2008)。

WPC80 经微滤处理去除脂肪,蛋白质浓缩至 90%,称为 WPI。WPI 有优良的功能性,包括胶凝性、乳化性和发泡性(Berry et al.,2009;Gaonkar et al.,2010)。

WPC80 和 WPI 也用作运动员和成人蛋白补充剂。

表 1.3　乳清产品组分　　　　　（单位：%）

	蛋白质	乳糖	脂肪	灰分	水
乳清粉	11.0~14.5	63.0~75.0	1.0~1.5	8.2~8.8	3.5~5.0
脱盐乳清	11.0~15.0	70.0~80.0	0.5~1.8	1.0~7.0	3.0~4.0
WPC34	34.0~36.0	48.0~52.0	3.0~4.5	6.5~8.0	3.0~4.5
WPC60	60.0~62.0	25.0~30.0	1.0~7.0	4.0~6.0	3.0~5.0
WPC80	80.0~82.0	4.0~8.0	4.0~8.0	3.0~4.0	3.5~4.5
WPI	90.0~92.0	0.5~1.0	0.5~1.0	2.0~3.0	4.5
滤液中固形物	3.0~8.0	65.0~85.0	<1.5	8.0~20.0	3.0~5.0

注：资料来自美国乳品出口委员会《美国乳清和乳糖产品参考手册》。

1.3.4　其他乳清蛋白质产品

不同的乳清蛋白质具有不同的功能和营养特性（Marshall，2004）。乳清蛋白质组分的分级分离将在第 2 章中讨论。α-乳白蛋白、β-乳球蛋白、乳铁蛋白、乳过氧化物酶和 GMP 等产品可在一些特殊和高端产品中应用。

1.3.5　乳矿物质

目前，各种乳清成分（包括乳矿物质）都得到了充分利用，如将 WPC 和 WPI 的加工副产物滤液中固形物喷成粉，或先将滤液中的乳糖通过结晶方法析出，再喷雾干燥得到主要成分为矿物质的乳盐（主要成分是乳糖和灰分，表 1.3）。滤液和乳盐都是良好的矿物质来源，可用作饲料和食品添加剂。

1.4　总　　结

乳清是牛奶由凝乳酶或酸凝固后而获得的黄色至绿色的透明液体，其利用已有几千年的历史，可用于饮料和其他食品。随着干酪制造业中乳清产量的增加和技术的发展，学术界和工业界都对乳清的利用做了大量的研究。乳糖、乳清蛋白质和乳盐等产品已经得到广泛应用。

参 考 文 献

Atra, R., Vatai, G., Bekassy-Molnar, E. et al. (2005). Investigation of ultra- and nanofiltration for utilization of whey protein and lactose. *Journal of Food Engineering* **67** (3): 325-332.

Beja-Pereira, A., Caramelli, D., Lalueza-Fox, C. et al. (2006). The origin of European cattle: evidence from modern and ancient DNA. *Proceedings of the National Academy of Sciences of the United States of America* **103** (21): 8113-8118.

Berry, R.A. (1923). The production, composition and utilisation of whey. *The Journal of Agricultural Science* **13** (2): 192-239.

Berry, T.K., Yang, X., and Foegeding, E.A. (2009). Foams prepared from whey protein isolate and egg white protein: 2. Changes associated with angel food cake functionality. *Journal of Food Science* **74** (5): E269-E277.

Briggs, H. (2012). Evidence of world's 'oldest' cheese-making found. Retrieved Dec 25, 2017, from www.bbc.co.uk/food/0/20695015 (accessed 12 Dec. 2012).

Brody, E.P. (2000). Biological activities of bovine glycomacropeptide. *British Journal of Nutrition* **84** (S1): 39-46.

Clutton-Brock, J. (1999). *A Natural History of Domesticated Mammals*. Cambridge: Cambridge University Press.

Dattatreya, A., Etzel, M.R., and Rankin, S.A. (2007). Kinetics of browning during accelerated storage of sweet whey powder and prediction of its shelf life. *International Dairy Journal* **17** (2): 177-182.

Daviau, C., Famelart, M.H., Pierre, A. et al. (2000). Rennet coagulation of skim milk and curd drainage: effect of pH, casein concentration, ionic strength and heat treatment. *Dairy Science and Technology* **80** (4): 397-415.

Dragone, G., Mussatto, S.I., Oliveira, J.M. et al. (2009). Characterisation of volatile compounds in an alcoholic beverage produced by whey fermentation. *Food Chemistry* **112** (4): 929-935.

Evershed, R.P., Payne, S., Sherratt, A.G. et al. (2008). Earliest date for milk use in the near east and southeastern Europe linked to cattle herding. *Nature* **455**: 528-531.

Francis, L.H. (1969). Process of treating whey. U.S. Patent 3, 447, 930, filed 14 April 1966 and issued 3 June 1969.

Gaonkar, G., Koka, R., Chen, K. et al. (2010). Emulsifying functionality of enzyme-modified milk proteins in O/W and mayonnaise-like emulsions. *African Journal of Food Science* **4** (1): 16-25.

Golding, J. and Rowsell, E. (1932). An auxiliary evaporating and preheating apparatus for drying whey, milk and other liquids by the roller process. *Journal of Dairy Research* **3** (3): 264-271.

Heffernan, C. (2015). A potted history of whey protein. https://physicalculturestudy.com/2015/12/07/a-potted-history-of-whey-protein (accessed 26 december 2015).

Holsinger, V.H. (1978). *Fortification of Soft Drinks with Protein from Cottage Cheese Whey*. New York and London: Plenum Press.

Holsinger, V.H. (1988). Lactose. In: *Fundamentals of Dairy Chemistry* (ed. N.P. Wong), 279-342. Berlin: Springer.

Hoppe, C., Andersen, G.S., Jacobsen, S. et al. (2008). The use of whey or skimmed milk powder in fortified blended foods for vulnerable groups. *The Journal of Nutrition* **138** (1): 145S-161S.

Houldsworth, D. (1980). Demineralization of whey by means of ion exchange and electrodialysis. *International Journal of Dairy Technology* **33** (2): 45-51.

Jost, R., Maire, J.C., Maynard, F. et al. (1999). Aspects of whey protein usage in infant nutrition, a brief review. *International Journal of Food Science and Technology* **34**: 533-542.

de Kruif, C.G. (1997). Skim milk acidification. *Journal of Colloid and Interface Science* **185** (1): 19-25.

de Kruif, C.G. and Holt, C. (2003). Casein micelle structure, functions and interactions. In: *Advanced Dairy Chemistry—1 Proteins* (ed. F. Fox and P.L.H. McSweeney), 233-276. Berlin: Springer.

Le Graët, Y. and Gaucheron, F. (1999). pH-induced solubilization of minerals from casein micelles: influence of casein concentration and ionic strength. *Journal of Dairy Research* **66** (2): 215-224.

Lu, C., Wang, G., Li, Y. et al. (2013). Effects of homogenisation pressures on physicochemical changes in different layers of ultra-high temperature whole milk during storage. *International Journal of Dairy Technology* **66** (3): 325-332.

Lucey, J. (2002). Formation and physical properties of milk protein gels. *Journal of Dairy Science* **85** (2): 281-294.

Lucey, J.A. (2003). Acid coagulation of milk. In: *Advanced Dairy Chemistry—1 Proteins* (ed. F. Fox and P.L.H. McSweeney), 1001-1025. Berlin: Springer.

Marshall, K. (2004). Therapeutic applications of whey protein. *Alternative Medicine Review* **9** (2): 136-156.

Mawson, A. (1994). Bioconversions for whey utilization and waste abatement. *Bioresource Technology* **47** (3): 195-203.

Merrell, L.C. (1911). Food obtained from whey. US Patent 985, 271, filed 27 July 1908 and issued 28 Feberary 1911.

Morr, C.V. and Ha, E.Y.W. (1993). Whey protein concentrates and isolates: processing and functional properties. *Critical Reviews in Food Science and Nutrition* **33** (6): 431-476.

Mycek, S. (n.d.) Alpine Therapy: The Whey Bath. Retrieved from http://www.insidersguidetospas.com/features/alpine-therapy-the-whey-bath.

O'Callaghan, D.J., O'Donnell, C.P., and Payne, F.A. (2002). Review of systems for monitoring curd setting during cheesemaking. *International Journal of Dairy Technology* **55** (2): 65-74.

Peebles, D.D. and Manning, P.D. (1939). Process of manufacturing a noncaking dried whey powder. US Patent 2, 181, 146, filed 9 December 1936 and issued 28 November 1939.

Penna, A.L.B., Baruffaldi, R., and Oliveira, M.N. (1997). Optimization of yogurt production using demineralized whey. *Journal of Food Science* **62** (4): 846-850.

Pescuma, M., Hébert, E.M., Mozzi, F. et al. (2010). Functional fermented whey-based beverage using lactic acid bacteria. *International Journal of Food Microbiology* **141** (1): 73-81.

Salque, M., Bogucki, P.I., Pyzel, J. et al. (2013). Earliest evidence for cheese making in the sixth millennium BC in northern Europe. *Nature* **493** (7433): 522-525.

Scott, R., Robinson, R.K., and Wilbey, R.A. (1998). *Cheesemaking Practice*. Berlin: Springer.

Singh, A.K. and Singh, K. (2012). Utilization of whey for the production of instant energy beverage by using response surface methodology. *Advance Journal of Food Science and Technology* **4** (2): 103-111.

Smithers, G.W. (2008). Whey and whey proteins-from 'gutter-to-gold'. *International Dairy Journal* **18** (7): 695-704.

Smithers, G.W. (2015). Whey-ing up the options-yesterday, today and tomorrow. *International Dairy Journal* **48**: 2-14.

Susli, H. (1956). New type of whey utilization: a lactomineral table beverage. Proceedings of the 14 th International Dairy Congress Volume l (Pt. 2) 477.

Tratnik, L. and Krsev, L. (1987). *Production of Fermented Beverage from Milk with Demineralized Whey*. Food and Agriculture Origanization of the United Nations.

Troy, C.S., MacHugh, D.E., Bailey, J.F. et al. (2001). Genetic evidence for near-eastern origins of European cattle. *Nature* **410**: 1088.

Tunick, M.H. (2008). Whey protein production and utilization: a brief history. In: *Whey Processing, Functionality and Health Benefits* (ed. C.I. Onwulata and P.J. Huth), 1-13. NJ: Wiley-Blackwell.

Webb, B.H. and Whittier, E.O. (1948). The utilization of whey: a review. *Journal of Dairy Science* **31** (2): 139-164.

de Wit, J.N. (1998). Nutritional and functional characteristics of whey proteins in food products. *Journal of Dairy Science* **81** (3): 597-608.

Yang, Y., Shevchenko, A., Knaust, A. et al. (2014). Proteomics evidence for kefir dairy in Early Bronze Age China. *Journal of Archaeological Science* **45**: 178-186.

Zydney, A.L. (1998). Protein separations using membrane filtration: new opportunities for whey fractionation. *International Dairy Journal* **8** (3): 243-250.

第2章 乳清蛋白质产品的生产工艺技术

Guorong Wang[1] and Mingruo Guo[1,2]

1. Department of Nutrition and Food Science, University of Vermont, Burlington, USA
2. College of Food Science, Northeast Agriculture University, Harbin, People's Republic of China

从干酪或其他凝乳中分离出来的乳清或其他乳清产品是一种稀液体，含有3.3%～6.0%乳糖、0.32%～0.7%蛋白质和 0.5%～0.7%矿物质（Marwaha and Kennedy，1998）。液体乳清中的主要干物质是乳糖、灰分和少量蛋白质，只有在经过浓缩和纯化之后，乳清蛋白质卓越的功能性质才能体现，乳清的利用程度取决于乳清蛋白质从液态乳清中回收、浓缩和纯化的程度。早期没有回收和浓缩蛋白质的技术，乳清只在动物饲料、土地肥料等方面有低值应用，喷雾干燥和膜过滤技术的应用使乳清蛋白质产业迅猛发展，这些技术在提取乳清蛋白质的同时能使蛋白质分子变性程度降到最低。此外，乳清蛋白质是一组具有不同特性、功能和营养组成的混合蛋白，其进一步分级分离（如制备富含β-乳球蛋白或α-乳白蛋白的WPC）将提高乳清蛋白质应用价值。本章将综述乳清蛋白质回收技术、乳清蛋白质组分的分离以及乳清产品加工。

2.1 乳清蛋白质回收技术

乳清蛋白质的生产是从液态乳清中去除乳糖、矿物质和脂肪，回收蛋白质并将蛋白质浓缩至一定纯度的过程。为实现这一目标，二十世纪进行了大量研究，包括选择性沉淀、膜技术、离子交换、电渗析和色谱技术等。乳清蛋白质对加工条件尤其是热非常敏感，因此加工对蛋白质分子结构产生的影响会决定乳清蛋白质产品质量和功能特性。

2.1.1 热/酸沉淀

乳清蛋白质对热敏感，加热后会发生热变性，这是一种结构"展开"的过程，可暴露出游离的巯基（SH），暴露的巯基引发巯基-二硫键（SH/SS）转换，形成初级胶体聚集体，很多因素（如钙和pH）能诱导进一步的非特异性聚集。许多学者（Donovan and Mulvihill，1987；Foegeding et al.，2002；Morr and Ha，1993；Mulvihill and Donovan，1987）对乳清蛋白质的热聚集或沉淀进行了详细的论述。乳清蛋白质的沉淀回收率取决于体系的 pH、温度和离子强度。与其他蛋白质不同的是，乳清蛋白质即使在等电点为 4.6 时，也有很强的酸性稳定性。等电点区（pH 4~5），加热和加钙有利于蛋白质沉淀回收，这一过程通常称为"热/酸沉淀"（Hill et al.，1982a，b）。

首先，将凝乳中分离出来的液体乳清通过真空蒸发或膜过滤（反向渗透或纳滤）浓缩增加干物质含量。Hill 等（1982a，b）发现 90℃及中性 pH（6.0~7.0）条件下乳清变性，然后调 pH 至 4.4~5.0 使之沉淀，可获得较高的蛋白质回收率，但此方法回收的乳清蛋白质的溶解性非常差（Mulvihill and Donovan，1987），而溶解性对于蛋白质的功能性质很重要。中性 pH 下加热乳清蛋白质比在酸性 pH 时加热产生更多的不可逆和不溶的蛋白聚集体（Mulvihill and Donovan，1987）。因此，为了提高乳清蛋白质的溶解度，酸化后加热可能比加热后再酸化更好，且添加铁可以提高乳清蛋白质的溶解度（Dalan et al.，1975）。酸化过程可以采用无机酸或有机酸，如柠檬酸，加热经酸化浓缩的甜乳清即可从液体乳清中沉淀并回收乳清蛋白质（Hill et al.，1982a，b）。研究发现，用柠檬酸将甜乳清 pH 调至 5.5 时，可获得较高的乳清蛋白质产量（Hill et al.，1982a，b）。Modler 和 Emmons（1977）通过热/酸沉淀法得到了高溶解度的 WPC（总蛋白含量为 60%~80%）。Molder 方法是将甜乳清的 pH 调至 2.5~3.5，然后在 90℃下加热 15 min，待溶液冷却后，将 pH 提高到等电点 4.5，再离心沉淀蛋白，最后将 pH 调回 6.0，以溶解沉淀物并喷雾干燥。Molder 和 Emmons（1977）还发现，铁的添加显著提高了 WPC 总蛋白含量（从 35%~53%增至 63%~74%），并且在加热前调 pH 至 2.5 所得的蛋白质溶解度远低于 pH 为 3.5 时得到的蛋白质的溶解度。

在膜技术广泛应用之前，热/酸沉淀是回收乳清蛋白质的常用方法，其优点是成本低，缺点是蛋白质纯度低、功能性差，酸化与中和反应增加了灰分含量，影响乳清蛋白质的应用。

2.1.2 膜加工技术

膜加工技术在乳品工业的应用已有半个多世纪，目前在乳清蛋白质生产中仍处于主导地位。膜加工技术的基本原理如图 2.1 所示。小于膜孔径的组分称为"渗透物"，粒径较大的被截留的组分称为"截留物"，因此膜可以分离不同粒径的组分。过滤系统有垂直和横流两种类型，如图 2.2 所示。垂直模式的进料方向直接朝向过滤膜，当渗透物穿过滤膜时，截留物颗粒可能沉积在过滤膜的表面导致膜快速污损（Youm et al.，1996）。横流模式下进料流动方向与滤膜方向相切，当截留物开始在膜的内表面积聚时，进料液流会将截留物从膜上清除，显著增加膜通量（Youm et al.，1996）。目前，乳清蛋白质生产多采用横流过滤技术，而垂直过滤仅限于小型的实验室操作。

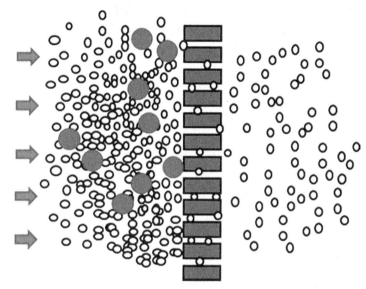

图 2.1　膜滤分离示意图

小于膜孔径的颗粒渗透过去，大于膜孔径的颗粒被保留

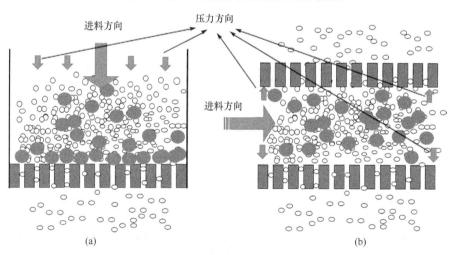

图 2.2　垂直过滤(a)与横流过滤(b)

牛奶含有粒径大小不一的颗粒，图 2.3 给出了牛奶中主要成分粒径大小以及不同膜的过滤范围。根据膜的孔径，膜滤过程可分为微滤（MF）、超滤（UF）、纳滤（NF）和反渗透（RO）。MF 能保留酪蛋白胶束、脂肪球和大于 100 nm 的细菌；UF 保留了脂肪球，以及大于 1 nm 的酪蛋白微粒和乳清蛋白质；NF 保留了

所有的蛋白质、脂肪和乳糖,只有矿物质和水可以通过膜;RO 保留了除自由水分子以外的所有物质(Kumar et al., 2013)。由于渗透物和截留物之间有浓度差异,因此需要外部压力来克服渗透压和控制流量。滤膜的孔径越小,则需要越高的压力来驱动渗透物穿过滤膜。MF、UF、NF 和 RO 所需的压力分别为 2 bar[①]、1~10 bar、5~40 bar 和 10~100 bar (Kumar et al., 2013)。

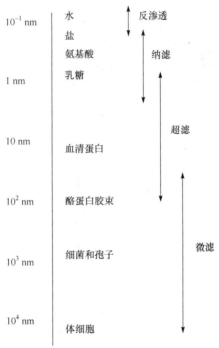

图 2.3　牛奶中主要成分的粒径大小和不同膜的过滤范围

目前,在 WPC 和 WPI 加工中,采用 UF 从液体乳清中回收乳清蛋白质,保留乳清蛋白质的同时可滤除乳糖和矿物质,因此蛋白质纯度在 20%~90%或 90%以上的、具有不同功能和用途的乳清蛋白质产品大多采用膜技术生产。WPC 和 WPI 的制备细节将在后面讨论,与热/酸沉淀相比,膜技术具有以下优点。

(1)非加热加工方法将乳清蛋白质的热破坏程度降至最低。

(2)不涉及蛋白质的相变化,乳清蛋白质在喷雾干燥前一直处于溶解状态。

(3)最少使用或不使用化学试剂,避免乳清蛋白质成品中出现高水平灰分或化学异味。

(4)节约成本。

(5)是一项清洁环保技术。

---

① 1 bar = 10^5 Pa。

2.2 乳清蛋白质组分的分离

乳清蛋白质是一组具有高营养价值的球状蛋白，每种组分都具有特定的营养作用或功能特性，乳清蛋白质组分的营养功能如表 2.1 所示，其营养价值将在第 5 章中介绍。乳清蛋白质整体上是营养价值高、功能性好的优质蛋白质（Bulut Solak and Akın，2012；de Wit，1998；Foegeding et al.，1992），但为了满足许多食品和营养品的特定要求，分离纯化乳清蛋白质各种组分的需求越来越高。

表 2.1　乳清蛋白质主要成分的物理化学特性和营养功能

乳清蛋白质组分	在乳清蛋白质中的占比/%	等电点（Wong et al.，1996；Zydney，1998）	摩尔质量（g/mol）（Wong et al.，1996；Zydney，1998）	营养功能
β-乳球蛋白（β-LG）	50~55	5.2	18362	必需氨基酸和支链氨基酸的来源（Hulmi et al.，2010）；增加维生素 A 摄入量（Said et al.，1989）；胶凝剂（Foegeding et al.，1992）
α-乳白蛋白（α-LA）	20~25	4.5~4.8	14147	人乳中的主要蛋白质（Heine et al.，1991）；必需氨基酸和支链氨基酸的来源；Ca 结合蛋白（Stuart et al.，1986）；非胶凝蛋白
牛血清白蛋白（BSA）	5~10	4.7~4.9	69000	必需氨基酸的来源
甜乳清中的糖巨肽（GMP）	10~15	—	7000	支链氨基酸的来源；补充芳香族氨基酸（用于苯丙酮尿症患者饮食）（Ney et al.，2009）
免疫球蛋白（Ig）	10~15	5.5~8.3	150000~1000000	"被动免疫"（Lyerly et al.，1991）
乳铁蛋白	1~2	9.0	78000	杀菌蛋白（Dionysius and Milne，1997）；铁结合特性（Nagasako et al.，1993）；肿瘤抑制剂（Iigo et al.，1999）
乳过氧化物酶	0.50	9.5	89000	抗菌活性（Björck et al.，1975）
溶菌酶	0.002	10.5~11	15000	抗菌活性（Jauregi and Welderufael，2010）

2.2.1　α-乳白蛋白和β-乳球蛋白的分离

α-乳白蛋白和β-乳球蛋白是乳清中的两种主要蛋白质，占乳清蛋白质总量的

70%以上。但α-乳白蛋白和β-乳球蛋白具有截然不同的营养和功能特性，α-乳白蛋白是婴儿营养的优质标准蛋白质，因为它是母乳中的主要蛋白质。牛乳中的α-乳白蛋白与人乳中的α-乳白蛋白具有 72%的氨基酸序列同源性（Heine et al., 1991），β-乳球蛋白是牛乳清蛋白质中的主要成分，但在人乳中并不存在。β-乳球蛋白可能会使婴儿配方奶粉的氨基酸组成与母乳不同（Heine et al., 1991），因此被认为是变应原（Ehn et al., 2004）。β-乳球蛋白具有良好的功能特性，如凝胶性、乳化性和发泡性（Rullier et al., 2008; Shimizu et al., 1985; Xiong et al., 1993），尤其是改性后其在酸奶、沙拉酱和代替鸡蛋等方面具有广泛应用（Guzey et al., 2004; Lee et al., 1994）。

α-乳白蛋白和β-乳球蛋白的分离最早可追溯到二十世纪五十年代（Aschaffenburg and Drewry, 1957; McKenzie, 1967）。实验室和工厂常使用选择性沉淀法来分离α-乳白蛋白和β-乳球蛋白。Na_2SO_4存在下乳清溶液的 pH 为 2 时，α-乳白蛋白沉淀而β-乳球蛋白仍处于溶解状态（Aschaffenburg and Drewry, 1957），后来，此方法改进为用$(NH_4)_2SO_4$代替Na_2SO_4（Armstrong et al., 1967）。早期化学沉淀法的明显缺点是引入了额外的矿物质却不能去除，研究仅局限于实验室规模。Amundson 等（1982）开发了一种中试规模的试验方法来分离α-乳白蛋白和β-乳球蛋白。该技术的流程图如图 2.4 所示：使用超滤技术将干酪乳清（切达干酪，pH 5.8）浓缩至体积减少 60%，用 HCl 或 NaOH 溶液将浓缩液的 pH 调至 4.65，再通过电渗析除盐，将脱盐后的浓缩液 pH 调至 4.65，再离心得到β-乳球蛋白沉淀，而α-乳白蛋白溶解在上清液中。β-乳球蛋白具有良好的持水性和乳化性，α-乳白蛋白具有更强的起泡特性（Amundson et al., 1982）。

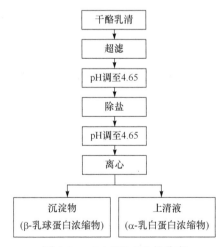

图 2.4　α-LA 和β-LG 的分离

改编自 Amundson et al.（1982）

后来研究发现，在接近等电点下选择性沉淀α-乳白蛋白比沉淀β-乳球蛋白更容易，Gésan-Guiziou 等（1999）改进了方法，用柠檬酸（而不是 HCl）调整 pH 到等电点以沉淀α-乳白蛋白。柠檬酸能够螯合α-乳白蛋白结合的钙，导致蛋白质分子不稳定（Hendrix，Griko，and Privalov，2000）。通过离心沉淀浓缩了α-乳白蛋白，上清液中富集的β-乳球蛋白经膜过滤进一步分离浓缩和纯化，通过离心和膜过滤分离，两种蛋白质的纯度均有了显著提高（Gésan-Guiziou et al.，1999）。

当 pH 在 3.0~4.6 之间时，α-乳白蛋白、牛血清白蛋白和免疫球蛋白对热非常敏感。当用柠檬酸调节 pH 到 3.0~4.6 并螯合α-乳白蛋白中的钙时，能得到高纯度的可溶性β-乳球蛋白（Maubois et al.，1987）。Toro-Sierra 等（2013）在中试中采用选择性热聚合和膜分离方式优化了α-乳白蛋白和β-乳球蛋白的分离方法。Toro-Sierra 等的方法主要步骤如图 2.5 所示。该方法使用 WPI 粉制备浓度为 15%的溶液，用柠檬酸和柠檬酸钠调 pH 到 3.4，使用柠檬酸钠是为了最大程度地增加柠檬酸基团浓度以螯合α-乳白蛋白中的钙。然后将所得溶液加热到 50℃沉淀α-乳白蛋白，再进行 24 h 的"老化聚合"，让聚集的α-乳白蛋白粒径逐渐增大，聚合完成的溶液经 MF（渗滤模式）分离α-乳白蛋白（截留物）和β-乳球蛋白

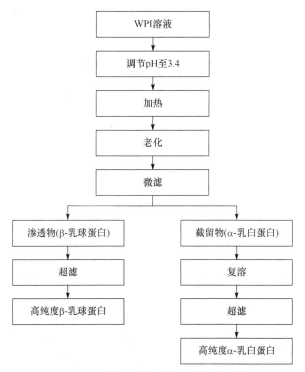

图 2.5 β-LG 和α-LA 的选择性热聚合和膜分离
改编自 Toro-Sierra et al.（2013）

(渗透物)，使用 NaOH 和 $CaCl_2$ 将 pH 调回至 8.0，让α-乳白蛋白的聚合物复溶，再通过 UF 进一步纯化蛋白质。将β-乳球蛋白溶液用水过滤清洗，通过 UF 将溶液中的柠檬酸和钙去除，然后用 NaOH 调节 pH 至 6.7。最终得到α-乳白蛋白纯度为 91.3%，产率为 60.7%~80.4%，β-乳球蛋白纯度为 97.2%，产率为 80.2%~97.3%。

乳清蛋白质溶液 pH 对于分离β-乳球蛋白和α-乳白蛋白至关重要，无机酸或盐酸调节 pH，引入了额外的灰分，需要在后期处理中去除以恢复蛋白质组分的功能。Bonnaillie 和 Tomasula（2012）开发了利用超临界二氧化碳作为酸化剂的新型蛋白质分离方法，如图 2.6 所示，将 WPI 溶液在高压反应器中加热到 60~65℃，然后将超临界 CO_2 泵入反应器，反应结束时，将温度冷却至 40℃，减压至大气压，减压乳清蛋白质溶液的最终 pH 约为 6.0，而未处理溶液的 pH 为 6.14~6.37。将样品在室温以 2000 g 离心 60 min，分别收集富集β-乳球蛋白的上清液和富集α-乳白蛋白的聚集体。上清液的蛋白质组成为 5%α-乳白蛋白、74%β-乳球蛋白和 18%~32%GMP，聚集体部分含有约 2%GMP、21%β-乳球蛋白和 62%α-乳白蛋白。该方法的优点是不会引入其他无机酸，通过 CO_2 将溶液酸化及通过释放 CO_2 来中和溶液。

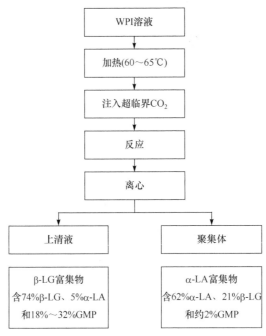

图 2.6 使用超临界 CO_2 分离β-LG 和α-LA
改编自 Bonnaillie and Tomasula（2012）

2.2.2 GMP 分离

天然酪蛋白分子表面覆盖着κ-酪蛋白的胶束结构，κ-酪蛋白亲水端伸入乳清相稳定胶束结构，伸出部分称为"κ-酪蛋白丝"，干酪生产中凝乳酶能够切断κ-酪蛋白丝，并将酪蛋白肽（残基106~169）释放到乳清中，称为酪蛋白糖巨肽（GMP或cGMP）。GMP不是乳清蛋白质，但它存在于甜乳清中，占乳清蛋白质总量的10%~15%。GMP缺乏芳香族氨基酸（苯丙氨酸、酪氨酸和色氨酸）或者芳香族氨基酸含量低，当含有GMP的乳清蛋白质应用于婴儿配方奶粉时，会增加配方粉与母乳氨基酸组成的差异。但GMP具有许多生物学活性，可用于特殊食品（Daddaoua et al., 2005; Ney et al., 2009），因此人们对含GMP的乳清蛋白质的需求正在日益增加，尤其是特殊婴儿配方奶粉的生产，同时市场也对GMP纯品感兴趣。

GMP相对于其他乳清蛋白质而言分子量更小，热稳定性更好。通过添加钙、调节pH和加热等方式，可以沉淀溶液中所有其他乳清蛋白质，保留GMP（Dosako et al., 1991），但加热至90℃以上会导致其他乳清蛋白质溶解度严重降低。

壳聚糖具有良好的阳离子特性，可用于吸附GMP。Li等（2010）发表了一种方法，使用一种改性壳聚糖从乳清中回收了90%左右的GMP。GMP分子粒径比其他乳清蛋白质小得多，所以膜过滤能有效分离出GMP，pH低于3时，GMP内部的唾液酸缺失，GMP变得不稳定，pH高于4时，GMP聚集并难以通过超滤膜（Tanimoto et al., 1991），将溶液pH调至3.5，然后超滤（分子质量20 kDa），能获得纯度约为80%的GMP，但产量很低（Tanimoto et al., 1991）。

离子交换色谱与超滤相结合可从乳清中回收GMP。Kawasaki和Dosako（1994）发明了一种用离子交换技术从乳清中分离GMP的方法，用阳离子交换剂（pH 3.0时使用磺酰基团）处理高达干酪乳清，搅拌培养20 h使其与GMP之外的乳清蛋白质结合，将含有GMP的滤液pH调至7.0，50℃下用截留分子质量为20000 Da的超滤膜进一步纯化，得到了纯度达到80%~88%的GMP。Xu等（2000）采用离子交换色谱和膜过滤方法分离出乳清中的免疫球蛋白G和GMP。

Bonnaillie和Tomasula的方法可让GMP存在于β-乳球蛋白中。Bonnaillie等（2014）开发的改进方法可进一步分离GMP和β-乳球蛋白，通过一条生产线便可获得三种蛋白质，其工艺流程图如图2.7所示，将反应釜中的WPI溶液加热至65~70℃以富集GMP，然后将CO_2泵入反应器，使pH达到4.9，以CO_2加压混合物反应305 min，冷却至40℃减压，将样品4000 g离心60 min，收集上清液（浓缩的β-乳球蛋白和GMP）和沉淀组分（浓缩的α-乳白蛋白）。将上清液超滤（截留分子质量为30 kDa）分离GMP和β-乳球蛋白，得到GMP溶液（浓度80.4%，回收率82%）和β-乳球蛋白浓缩物（59.9%β-乳球蛋白、27.1%GMP和13.0%α-乳白蛋白）。

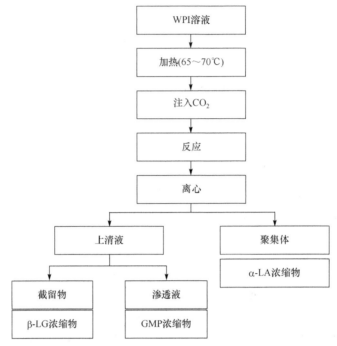

图 2.7 采用超临界 CO_2 和超滤分离 GMP、α-LA 和 β-LG（Bonnaillie et al., 2012）

2.2.3 牛血清白蛋白和免疫球蛋白分离

牛血清白蛋白（BSA）也是乳清中的主要蛋白质（中性），但一般不单独分离出来，作为乳清蛋白质产品中重要的氨基酸来源，不改变其他乳清蛋白质的整体性能。α-LA、β-LG 和 BSA 可以通过凝胶过滤和阴离子交换色谱进行分离（Neyestani et al., 2003）。

免疫球蛋白摩尔质量在乳清蛋白质中最大（表 2.1）。当 α-LA、β-LG 和 BSA 处于滤过液中时，利用 100000 或 500000 分子量截留膜可得到约 10%免疫球蛋白的截留物（Bottomley, 1993; Stott and Lucas, 1989）。免疫球蛋白的进一步纯化可通过阴离子交换系统实现（Xu et al., 2000），Xu 等（2000）通过使用聚苯乙烯阴离子交换剂和 UF（截留分子量为 100000）选择性地去除 α-LA、β-LG 和 BSA，浓缩免疫球蛋白，所得产物含有 43.3%的免疫球蛋白 G。

2.2.4 乳铁蛋白和乳过氧化物酶分离

乳铁蛋白和乳过氧化物酶的等电点分别为 9.0 和 9.5，与其他主要的蛋白质存在显著差异（表 2.1），据此可将两种蛋白质分离，常规实验室分离方法是沉淀技术和离子交换色谱法（Morrison et al., 1957; Morrison and Hultquist, 1963）。

经过几十年发展，阳离子交换色谱技术已能生产乳铁蛋白和乳过氧化物酶产品（Burling，1992；Chiu and Etzel，1997；Fuda et al.，2004；Yoshida and Ye，1991）。美国专利（No.5149647）公开了从甜乳清中提取纯乳铁蛋白和乳过氧化物酶的方法，甜乳清首先通过微滤去除脂肪和大的蛋白质聚集体，再通过快速流动型强阳离子交换床，乳过氧化物酶和乳铁蛋白分别用 0.3 mol/L NaCl 和 0.9 mol/L NaCl 的磷酸盐缓冲液洗脱（Burling，1992），将收集到的蛋白质组分脱盐并冷冻干燥。Andersson 和 Mattiasson（2006）使用模拟移动床连续色谱技术简化了分离 WPC 中乳过氧化物酶和乳铁蛋白的工艺。离子交换也可用于从甜乳清中分离乳铁蛋白和乳过氧化物酶以及其他乳清蛋白质，包括α-乳白蛋白和β-乳球蛋白（Ye et al.，2000），β-乳球蛋白用弱阴离子交换剂（diethylaminoethyl-Toyopearl）分离，α-乳白蛋白用强阳离子交换剂（quaternary aminoethyl-Toyopearl）分离，乳过氧化物酶和乳铁蛋白用强阳离子交换剂（sulphopropyl-Toyopearl）分离。市售乳铁蛋白主要是用离子交换技术生产，纯度达 90%以上，生产过程如图 2.8 所示。

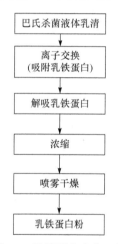

图 2.8　乳铁蛋白生产工艺

亲和色谱法可以提取高纯度乳铁蛋白和乳过氧化物酶。螯合色谱法用于分离干酪乳清中的免疫球蛋白和乳铁蛋白（Al-Mashikhi et al.，1988）。由于乳铁蛋白和肝素之间的特定结合能力，乳铁蛋白能够被微米级单分散超顺磁性聚甲基丙烯酸缩水甘油酯（PGMA）颗粒与肝素偶联（PGMA-肝素）吸附（Chen et al.，2007）。

胶态气泡（CGAs）是剧烈搅拌表面活性剂溶液产生的高比表面积微气泡。CGAs 通过静电和疏水作用可对分子产生特殊的吸附作用，可用于蛋白质的分离（Noble et al.，1998；O'Connell and Varley，2001）。Fuda 等（2004）开发了使用阴离子表面活性剂双-2-乙基己基磺基琥珀酸钠（AOT）生成的 CGAs 从甜乳清中回收乳铁蛋白和乳过氧化物酶的方法。α-乳白蛋白、β-乳球蛋白、BSA 的等电点在 4.5～5.5 左右，乳铁蛋白和乳过氧化物酶的等电点在 7～9.5 左右。当 pH 为 4.0

时，α-乳白蛋白、β-乳球蛋白和 BSA 的净电荷几乎为零或仅略带正电，而乳铁蛋白和乳过氧化物酶则带正电，AOT 产生的 CGAs 发挥了阳离子交换作用，选择性地吸附乳铁蛋白和乳过氧化物酶，从而将溶液分成微泡沫相（浓缩了乳铁蛋白和乳过氧化物酶）和液相，在微泡沫相的乳铁蛋白和乳过氧化物酶是液相的 25 倍，回收率为 90%。

2.3 乳清产品加工

乳清和乳清蛋白质产品主要是甜乳清粉、脱盐乳清粉、WPC、WPI 和其他乳清组分。这些产品的加工过程如图 2.9 所示。

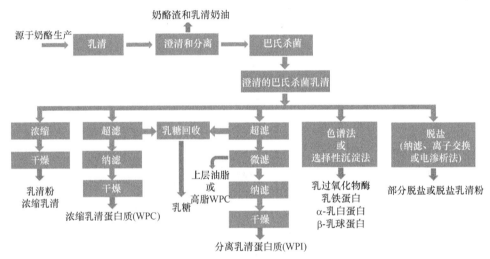

图 2.9　乳清主要成分及乳清蛋白质产品加工

2.3.1　液体乳清的澄清、分离和巴氏杀菌

目前，乳清蛋白质产品主要源自干酪和酪蛋白加工过程中产生的液体乳清。液体乳清主要有甜乳清和酸乳清两种类型。甜乳清是指通过不显著降低 pH 的加工（如切达干酪和马苏里拉干酪）产生的乳清，是全球主要的商品乳清。酸乳清则通常来自农家干酪和酸性酪蛋白的生产。二十一世纪初，美国和其他国家掀起了希腊酸奶的消费热潮（Gurel，2016），因而产生的酸乳清越来越多。甜乳清和酸乳清的主要差别是 pH 和矿物质。酸乳清 pH 的降低是由乳酸（如农家干酪乳清和希腊酸奶乳清）或无机酸（如酸性酪蛋白乳清）引起的，随着 pH 的降低，酪蛋白微粒中的钙以胶体磷酸钙的形式溶解（Dalgleish and Law，1989），导致乳清中钙和总灰分含量增加，如表 2.2 所示。酸乳清中不含由凝乳酶水解的κ-酪蛋白

片段 GMP，表明它适宜在婴儿食品中应用。但酸乳清比甜乳清处理起来更加困难，因为酸度高，导致喷雾干燥后的粉末更黏稠（Modler and Emmons，1978）。现在，乳酸可以通过纳滤和电渗析去除（Kelly and Kelly，1995；Román et al.，2009），使酸乳清和甜乳清一样有价值（Soehnlen，1982）。

表 2.2 甜乳清和酸乳清的组成

	甜乳清（切达干酪乳清）	酸乳清（农家干酪乳清）
pH	6.3	4.6
灰分	0.53	0.69
乳糖	4.77	0.71
蛋白质	0.82	0.75
脂肪	0.07	0.03
乳酸	0.15	0.55
钙	0.05	0.13
钠	0.07	0.06
钾	0.13	0.15
磷	0.06	0.09

注：资料改编自 Morr and Ha（1993）。

干酪工厂产生的液体乳清含有酪蛋白/干酪渣和脂肪，两者都会对后续工艺产生不利影响，因此应首先清除。酪蛋白/干酪渣可通过澄清步骤被除去，或采用如离心机、筛网和旋风分离器等设备去除细渣，去除的细渣常被添回到干酪生产中。澄清后，用离心分离机分离脂肪得到乳清奶油（25%～30%的脂肪），可再用于制干酪或制成含乳清黄油（Halpin-Dohnalek and Marth，1989）。澄清和分离的乳清必须经过巴氏杀菌并冷却至5℃以下，尽可能将其冷藏保存以备进一步处理。

2.3.2 膜过滤

膜过滤是生产 WPC 和 WPI 的下一个步骤。如果乳清需要运输到不同的地方进行处理，为了增加干物质的含量，需要采取浓缩步骤，如采用反渗透、反渗透结合纳滤或蒸发，以节省运输成本。WPC 和 WPI 的生产旨在去除液态乳清中的乳糖并浓缩蛋白质。超滤可用于分离乳糖和乳清蛋白质。通常会进行渗滤以增加膜通量以及蛋白质含量。无渗滤时，干基蛋白质含量最高仅为 60%～70%，而用渗滤则可提高到 80%以上。根据蛋白质浓度的要求，用超滤法获得的蛋白质含量在 20%～80%之间。超滤也可保留脂肪和蛋白质，并透过乳糖和矿物质，在浓缩乳清蛋白质的同时，成品粉中的脂肪也随着蛋白质一起浓缩（表 1.3）。WPC80

中的脂肪含量高达 4%～8%，因此无法进一步纯化蛋白质。为了将蛋白质进一步增加至 90%以制成 WPI，必须去除脂肪，可采用微滤技术。WPC 和 WPI 制作工艺如图 2.9 所示，截留物部分含有高浓度脂肪和变性乳清蛋白质，它们是 WPI 加工过程中的副产物，称为原奶油或高脂 WPC。许多研究人员对原奶油的营养、功能和应用进行了研究，以增加其商业价值（Bund and Hartel, 2013；Li et al., 2016；Sünder et al., 2001）。

2.3.3 脱盐

乳清粉中的灰分含量过高（8.2%～8.8%）限制了它的运用，尤其是在婴儿食品方面（Jost et al., 1999；Tunick, 2008）。灰分含量不超过 7%才能称为脱盐或低盐乳清粉。乳清粉的脱盐率一般为 25%、50%和 90%。离子交换、电渗析和纳滤是常用的脱盐方法（Houldsworth, 1980；van der Horst et al., 1995）。离子交换过程包括阳离子交换和阴离子交换，利用树脂珠去除溶液中的矿物，当树脂珠饱和后，排出脱盐溶液，树脂珠再生；电渗析是在直流电的驱动下，通过半透膜输送离子来达到分离离子的目的，所使用的膜具有阴离子和阳离子交换功能，能够降低所通过液体的盐含量。离子交换和电渗析法可提高脱盐率。纳滤膜只允许矿物质和水通过，能实现部分脱盐，但与离子交换和电渗析法相比，纳滤更适合大规模连续性生产。

2.3.4 浓缩

乳清粉含有 63%～75%的乳糖。像奶粉一样，如果乳糖在干燥前没有得到适当处理，成品粉中会出现吸湿性很强的非结晶乳糖，导致严重的货架期问题，如成团和结块（Ibach and Kind, 2007）。在乳清粉和低蛋白 WPC 生产中，需要使用真空蒸发器将溶液的干物质浓缩至 45%～65%，然后快速冷却至 30～40℃，在此条件下乳糖变成过饱和状态结晶析出（主要是不吸湿的α-乳糖晶体），浓缩需要保持 4～8 h 以获得小而均匀的乳糖晶体。高蛋白的 WPC 和 WPI 产品中乳糖含量较低，对其货架期的影响弱于乳清粉和全脂奶粉。此外，乳清蛋白质对热非常敏感，特别是在高浓度的溶液中。纳滤是一种冷处理技术，可最大限度地减少乳清蛋白质变性，所以在 WPC80 和 WPI 的生产中，常用纳滤法来代替真空蒸发，不仅可使干物质浓度由 20%提高到 30%～40%，还可部分脱盐，降低 WPC80 和 WPI 成品的灰分含量。

2.3.5 干燥

干燥是乳清和乳清粉加工的最后一个步骤，主要有滚筒干燥和喷雾干燥两种方法，冷冻干燥通常用于生产高纯度（即分析纯级）乳清组分，但不适用于食品级乳清蛋白质生产。甜乳清很可能会粘在滚筒式干燥设备的内表面上，难以刮除。喷雾干燥是乳清工业中最主要的干燥方法。

在喷雾干燥之前乳糖结晶对生产优质乳清粉至关重要，直接喷雾干燥，会产生极易吸湿的非结晶乳糖。为了提高乳清粉的分散性，进一步增加乳清粉中的乳糖晶体，通常还需要使用流化床。喷雾干燥对乳清蛋白质性质有影响，如溶解度、体积密度和其他功能（Schmidt et al., 1984）。WPC80 和 WPI 可在喷雾干燥时喷涂卵磷脂，提高其速溶性、分散性，使 WPC80 和 WPI 粉适用于运动营养配方中。

2.4 总　　结

乳清工业的发展主要依赖于乳清生产技术的进步。由于乳清蛋白质的功能特性和营养价值在高乳糖/灰分的乳清液和稀乳清中表现不佳，因此对乳清中的各种组分（特别是蛋白质）的分离和回收是关键。乳清用作动物饲料或田间肥料已有几千年的历史，喷雾干燥技术和蛋白膜分离技术的出现提高了乳清利用附加值。乳清蛋白质被认为是最重要的乳制品成分之一，膜分离技术是乳清加工的核心技术，膜技术的优点已被众多研究人员所重视，特别是适用于对热敏感的乳清蛋白质加工，可使乳清蛋白质尽可能保持天然优良的功能特性。乳清蛋白质是一组混合蛋白，不仅含有高营养蛋白组分，还含有数量虽少，但种类多样、分子量各异的生物活性蛋白，因此膜过滤法也可用于乳清组分的分级分离。离子交换、膜色谱、选择性沉淀等是进一步纯化乳清蛋白质的常用方法，特别是对痕量蛋白和微量蛋白的纯化。目前，乳铁蛋白和乳过氧化物酶等大部分乳清组分产品均已上市销售。

参 考 文 献

Al-Mashikhi, S.A., Li-Chan, E., and Nakai, S. (1988). Separation of immunoglobulins and lactoferrin from cheese whey by chelating chromatography. *Journal of Dairy Science* **71** (7): 1747-1755.

Amundson, C.H., Watanawanichakorn, S., and Hill, C.G. Jr. (1982). Production of enriched protein fractions of β-Lg and α-La from cheese whey. *Journal of Food Processing and Preservation* **6** (2): 55-71.

Andersson, J. and Mattiasson, B. (2006). Simulated moving bed technology with a simplified approach for protein purification: separation of lactoperoxidase and lactoferrin from whey protein concentrate. *Journal of Chromatography A* **1107** (1-2): 88-95.

Armstrong, J.M., McKenzie, H., and Sawyer, W.H. (1967). On the fractionation of β-lactoglobulin and α-lactalbumin. *Biochimica et Biophysica Acta (BBA)-Protein Structure* **147** (1): 60-72.

Aschaffenburg, R. and Drewry, J. (1957). Improved method for the preparation of crystalline β-lactoglobulin and α-lactalbumin from cow's milk. *Biochemical Journal* **65** (2): 273.

Björck, L., Rosen, C., Marshall, V. et al. (1975). Antibacterial activity of the lactoperoxidase system in milk against pseudomonads and other gram-negative bacteria. *Applied Microbiology* **30** (2): 199-204.

Bonnaillie, L.M. and Tomasula, P.M. (2012). Fractionation of whey protein isolate with supercritical carbon dioxide to produce enriched α-lactalbumin and β-lactoglobulin food ingredients. *Journal of Agricultural and Food Chemistry* **60** (20): 5257-5266.

Bottomley, R.C. (1993). Isolation of an immunoglobulin rich fracton from whey. US Patent 5,194,591, filed 5 December 1988 and issued 16 March 1993.

Bulut Solak, B. and Akin, N. (2012). Functionality of whey protein. *International Journal of Health and Nutrition* **3** (1): 1-7.

Bund, R.K. and Hartel, R.W. (2013). Blends of delactosed permeate and pro-cream in ice cream: effects on physical, textural and sensory attributes. *International Dairy Journal* **31** (2): 132-138.

Burling, H. (1992). Process for extracting pure fractions of lactoperoxidase and lactoferrin from milk serum. US Patent 5,149,647, filed 25 November 1988 and issued 22 September 1922.

Chen, L., Guo, C., Guan, Y.P. et al. (2007). Isolation of lactoferrin from acid whey by magnetic affinity separation. *Separation and Purification Technology* **56** (2): 168-174.

Chiu, C.K. and Etzel, M.R. (1997). Fractionation of lactoperoxidase and lactoferrin from bovine whey using a cation exchange membrane. *Journal of Food Science* **62** (5): 996-1000.

Daddaoua, A., Puerta, V., Zarzuelo, A. et al. (2005). Bovine glycomacropeptide is anti-inflammatory in rats with hapten-induced colitis. *The Journal of Nutrition* **135** (5): 1164-1170.

Dalan, E., Groux, M.J.A., Tour-de-Peilz, L. et al. (1975). Preparation of a soluble whey protein fraction. US Patent 3,922,375 A, filed 18 July 1973 and issued 25 November 1975.

Dalgleish, D.G. and Law, A.J. (1989). pH-induced dissociation of bovine casein micelles II. Mineral solubilization and its relation to casein release. *Journal of Dairy Research* **56** (05): 727-735.

Dionysius, D. and Milne, J. (1997). Antibacterial peptides of bovine lactoferrin: purification and characterization. *Journal of Dairy Science* **80** (4): 667-674.

Donovan, M. and Mulvihill, D.M. (1987). Thermal denaturation and aggregation of whey proteins. *Irish Journal of Food Science and Technology* **11** (1): 87-100.

Dosako, S., Nishiya, T. and Deya, E. (1991). Process for the production of κ-casein glycomacropeptide. US Patent 5,061,622, filed 6 May 1988 and issued 29 October 1991.

Ehn, B., Ekstrand, B., Bengtsson, U. et al. (2004). Modification of IgE binding during heat processing of the cow's milk allergen β-lactoglobulin. *Journal of Agricultural and Food Chemistry* **52** (5): 1398-1403.

Foegeding, E.A., Davis, J.P., Doucet, D. et al. (2002). Advances in modifying and understanding whey protein functionality. *Trends in Food Science and Technology* **13** (5): 151-159.

Foegeding, E.A., Kuhn, P.R., and Hardin, C.C. (1992). Specific divalent cation-induced changes during gelation of beta-lactoglobulin. *Journal of Agricultural and Food Chemistry* **40** (11): 2092-2097.

Fuda, E., Jauregi, P., and Pyle, D.L. (2004). Recovery of lactoferrin and lactoperoxidase from sweet whey using colloidal gas aphrons (CGAs) generated from an anionic surfactant, AOT.

Biotechnology Progress **20** (2): 514-525.

Gésan-guiziou, G., Daufin, G., Timmer, M. et al. (1999). Process steps for the preparation of purified fractions of α-lactalbumin and β-lactoglobulin from whey protein concentrates. *Journal of Dairy Research* **66** (02): 225-236.

Gurel, P. (2016). Live and active cultures: gender, ethnicity, and "Greek" yogurt in America. *Gastronomica: The Journal of Critical Food Studies* **16** (4): 66-77.

Guzey, D., Kim, H., and McClements, D.J. (2004). Factors influencing the production of o/w emulsions stabilized by β-lactoglobulin-pectin membranes. *Food Hydrocolloids* **18** (6): 967-975.

Halpin-Dohnalek, M.I. and Marth, E.H. (1989). Fate of *Staphylococcus aureus* in whey, whey cream, and whey cream butter. *Journal of Dairy Science* **72** (12): 3149-3155.

Heine, W.E., Klein, P.D., and Reeds, P.J. (1991). The importance of alpha-lactalbumin in infant nutrition. *The Journal of Nutrition* **121** (3): 277-283.

Hendrix, T., Griko, Y.V., and Privalov, P.L. (2000). A calorimetric study of the influence of calcium on the stability of bovine α-lactalbumin. *Biophysical Chemistry* **84** (1): 27-34.

Hill, A.R., Bullock, D.H., and Irvine, D.M. (1982a). Recovery of whey proteins from concentrated sweet whey. *Canadian Institute of Food Science and Technology Journal* **15** (3): 180-184.

Hill, A.R., Irvine, D.M., and Bullock, D.H. (1982b). Precipitation and recovery of whey proteins: a review. *Canadian Institute of Food Science and Technology Journal* **15** (3): 155-160.

van der Horst, H., Timmer, J., Robbertsen, T. et al. (1995). Use of nanofiltration for concentration and demineralization in the dairy industry: model for mass transport. *Journal of Membrane Science* **104** (3): 205-218.

Houldsworth, D. (1980). Demineralization of whey by means of ion exchange and electrodialysis. *International Journal of Dairy Technology* **33** (2): 45-51.

Hulmi, J.J., Lockwood, C.M., and Stout, J.R. (2010). Effect of protein/essential amino acids and resistance training on skeletal muscle hypertrophy: a case for whey protein. *Nutrition & Metabolism* **7** (1): 51-62.

Ibach, A. and Kind, M. (2007). Crystallization kinetics of amorphous lactose, whey-permeate and whey powders. *Carbohydrate Research* **342** (10): 1357-1365.

Iigo, M., Kuhara, T., Ushida, Y. et al. (1999). Inhibitory effects of bovine lactoferrin on colon carcinoma 26 lung metastasis in mice. *Clinical & Experimental Metastasis* **17** (1): 43-49.

Jauregi, P. and Welderufael, F.T. (2010). Added-value protein products from whey. *Nutrafoods* **9** (4): 13-23.

Jost, R., Maire, J.C., Maynard, F. et al. (1999). Aspects of whey protein usage in infant nutrition, a brief review. *International Journal of Food Science and Technology* **34** (5-6): 533-542.

Kawasaki, Y. and Dosako, S. (1994). Process for producing κ-casein glycomacropeptides. US Patent 5,278,288, filed 26 November 1991 and issued 11 January 1994.

Kelly, J. and Kelly, P. (1995). Desalination of acid casein whey by nanofiltration. *International Dairy Journal* **5** (3): 291-303.

Kumar, P., Sharma, N., Ranjan, R. et al. (2013). Perspective of membrane technology in dairy industry: a review. *Asian-Australasian Journal of Animal Sciences* **26** (9): 1347-1358.

Lee, S.P., Kim, D.S., Watkins, S. et al. (1994). Reducing whey syneresis in yogurt by the addition of a thermolabile variant of β-lactoglobulin. *Bioscience, Biotechnology, and Biochemistry* **58** (2): 309-313.

Li, B., Linghu, Z., Hussain, F. et al. (2016). Extraction of phospholipids from procream using supercritical carbon dioxide and ethanol as a modifier. *Journal of Animal Science* **94** (S5): 256-256.

Li, C., Song, X.H., Hein, S. et al. (2010). The separation of GMP from milk whey using the modified chitosan beads. *Adsorption* **16** (1): 85-91.

Lyerly, D.M., Bostwick, E., Binion, S.B. et al. (1991). Passive immunization of hamsters against disease caused by Clostridium difficile by use of bovine immunoglobulin G concentrate. *Infection and Immunity* **59** (6): 2215-2218.

Marwaha, S.S. and Kennedy, J.F. (1988). Whey-pollution problem and potential utilization. *International Journal of Food Science & Technology* **23** (4): 323-336.

Maubois, J., Pierre, A., Fauquant, J. et al. (1987). *Industrial Fractionation of Main Whey Proteins*. Food and Agriculture Organization of the United Nations.

McKenzie, H.A. (1967). Milk proteins. In: *Advances in Protein Chemistry* (ed. C.B. Anfinsen Jr., A.L. Anson, J.T. Edsall and F.M. Richards), 55-234. MA: Academic Press.

Modler, H.W. and Emmons, D.B. (1977). Properties of whey protein concentrate prepared by heating under acidic conditions. *Journal of Dairy Science* **60** (2): 177-184.

Modler, H.W. and Emmons, D.B. (1978). Calcium as an adjuvant for spray-drying acid whey. *Journal of Dairy Science* **61** (3): 294-299.

Morr, C.V. and Ha, E.Y.W. (1993). Whey protein concentrates and isolates: processing and functional properties. *Critical Reviews in Food Science and Nutrition* **33** (6): 431-476.

Morrison, M., Hamilton, H.B., and Stotz, E. (1957). The isolation and purification of lactoperoxidase by ion exchange chromatography. *Journal of Biological Chemistry* **228** (2): 767-776.

Morrison, M. and Hultquist, D.E. (1963). Lactoperoxidase: II isolation. *Journal of Biological Chemistry* **238** (8): 2847-2849.

Mulvihill, D. and Donovan, M. (1987). Whey proteins and their thermal denaturation - a review. *Irish Journal of Food Science and Technology* **11** (1): 43-75.

Nagasako, Y., Saito, H., Tamura, Y. et al. (1993). Iron-binding properties of bovine lactoferrin in ironrich solution. *Journal of Dairy Science* **76** (7): 1876-1881.

Ney, D.M., Gleason, S.T., van Calcar, S.C. et al. (2009). Nutritional management of PKU with glycomacropeptide from cheese whey. *Journal of Inherited Metabolic Disease* **32** (1): 32-39.

Neyestani, T.R., Djalali, M., and Pezeshki, M. (2003). Isolation of α-lactalbumin, β-lactoglobulin, and bovine serum albumin from cow's milk using gel filtration and anion-exchange chromatography including evaluation of their antigenicity. *Protein Expression and Purification* **29** (2): 202-208.

Noble, M., Brown, A., Jauregi, P. et al. (1998). Protein recovery using gas-liquid dispersions. *Journal of Chromatography B: Biomedical Sciences and Applications* **711** (1): 31-43.

O'Connell, P. and Varley, J. (2001). Immobilization of Candida rugosa lipase on colloidal gas aphrons

(CGAs). *Biotechnology and Bioengineering* **74** (3): 264-269.
Román, A., Wang, J., Csanadi, J. et al. (2009). Partial demineralization and concentration of acid whey by nanofiltration combined with diafiltration. *Desalination* **241** (1): 288-295.
Rullier, B., Novales, B., and Axelos, M.A.V. (2008). Effect of protein aggregates on foaming roperties of β-lactoglobulin. *Colloids and Surfaces A: Physicochemical and Engineering Aspects* **330** (2): 96-102.
Said, H.M., Ong, D.E., and Shingleton, J.L. (1989). Intestinal uptake of retinol: enhancement by bovine milk beta-lactoglobulin. *The American Journal of Clinical Nutrition* **49** (4): 690-694.
Schmidt, R.H., Packard, V.S., and Morris, H.A. (1984). Effect of processing on whey protein functionality. *Journal of Dairy Science* **67** (11): 2723-2733.
Shimizu, M., Saito, M., and Yamauchi, K. (1985). Emulsifying and structural properties of β-lactoglobulin at different pHs. *Agricultural and Biological Chemistry* **49** (1): 189-194.
Soehnlen, J.A. (1982). Process for converting sour whey into sweet whey and product. US Patent 4,358,464, filed 14 August 1978 and issued 9 November 1982.
Stott, G.H. and Lucas, D.O. (1989). Immunologically active whey fraction and recovery process. US Patent 4,834,974, 24 December 1986 and issued 30 May 1989.
Stuart, D., Acharya, K., Walker, N.P.C. et al. (1986). α-Lactalbumin possesses a novel calcium binding loop. *Nature* **324** (6092): 84-87.
Sünder, A., Scherze, I., and Muschiolik, G. (2001). Physico-chemical characteristics of oil-in-water emulsions based on whey protein-phospholipid mixtures. *Colloids and Surfaces B: Biointerfaces* **21** (1): 75-85.
Tanimoto, M., Kawasaki, Y., Shinmoto, H., Dosako, S., and Tomizawa, A. (1991). Process for producing κ-casein glycomacropeptides. US Patent 5,075,424 A, filed 26 November 1991 and issued 11 January 1994.
Toro-Sierra, J., Tolkach, A., and Kulozik, U. (2013). Fractionation of α-lactalbumin and β-lactoglobulin from whey protein isolate using selective thermal aggregation, an optimized membrane separation procedure and resolubilization techniques at pilot plant scale. *Food and Bioprocess Technology* **6** (4): 1032-1043.
Tunick, M.H. (2008). Whey protein production and utilization: a brief history. In: *Whey Processing, Functionality and Health Benefits* (ed. C. Onwulata and P. Huth), 1-13. Berlin: Springer.
de Wit, J. (1998). Nutritional and functional characteristics of whey proteins in food products. *Journal of Dairy Science* **81** (3): 597-608.
Wong, D.W., Camirand, W.M., Pavlath, A.E. et al. (1996). Structures and functionalities of milk proteins. *Critical Reviews in Food Science and Nutrition* **36** (8): 807-844.
Xiong, Y.L., Dawson, K.A., and Wan, L.P. (1993). Thermal aggregation of β-lactoglobulin: effect of pH, ionic environment, and thiol reagent. *Journal of Dairy Science* **76** (1): 70-77.
Xu, Y., Sleigh, R., Hourigan, J. et al. (2000). Separation of bovine immunoglobulin G and glycomacropeptide from dairy whey. *Process Biochemistry* **36** (5): 393-399.
Ye, X., Yoshida, S., and Ng, T.B. (2000). Isolation of lactoperoxidase, lactoferrin, α-lactalbumin, β-lactoglobulin B and β-lactoglobulin A from bovine rennet whey using ion exchange

chromatography. *The International Journal of Biochemistry & Cell Biology* **32** (11-12): 1143-1150.

Yoshida, S. and Ye, X. (1991). Isolation of lactoperoxidase and lactoferrins from bovine milk acid whey by carboxymethyl cation exchange chromatography. *Journal of Dairy Science* **74** (5): 1439-1444.

Youm, K.H., Fane, A.G., and Wiley, D.E. (1996). Effects of natural convection instability on membrane performance in dead-end and cross-flow ultrafiltration. *Journal of Membrane Science* **116** (2): 229-241.

Zydney, A.L. (1998). Protein separations using membrane filtration: new opportunities for whey fractionation. *International Dairy Journal* **8** (3): 243-250.

第 3 章 乳清蛋白质的化学特性

Mingruo Guo[1,2] and Cuina Wang[3]

1. Department of Nutrition and Food Science, University of Vermont, Burlington, USA
2. College of Food Science, Northeast Agriculture University, Harbin, People's Republic of China
3. Department of Food Science, College of Food Science and Engineering, Jilin University, ChangChun, People's Republic of China

乳清蛋白质是多种分泌蛋白的混合物。它的主要成分包括β-乳球蛋白（β-LG）、α-乳白蛋白（α-LA）、牛血清白蛋白（BSA）和一些其他的蛋白质，如乳铁蛋白（Lf）、免疫球蛋白（Ig）和微量蛋白质等（Smithers et al.，1996）（表3.1）。

表 3.1 主要乳清蛋白质的化学特性

	β-LG	α-LA	BSA	Lf	Ig
含量（%）	50	20	10	＜ 0.1	3
氨基酸残基（个）	162	123	583	689	＞500
分子质量（kDa）	18.4	14.2	64	76.1	16.1
变性温度（℃）	70～73	64	55～85	—	—
SH/SS	1/2	0/4	1/17	—	—
钙离子含量（个）	0	1	—	—	—
等电点	5.2/5.4	4.4	5.5	—	5.1～8.3
变体种类	2	1	2	—	2（分别位于轻链和重链）

注：资料来自 Aoki et al.（1969）、Farrell Jr. et al.（2004）、Hoffmann et al.（1997）、Boye and Alli（2000）、Phillips and Williams（2011）等参考文献。

3.1 β-乳球蛋白

3.1.1 β-乳球蛋白的化学特性

β-LG 约占牛乳总乳清蛋白质的 50%～60%，是乳清的主要成分（Liang et al.，2016）。1934 年首次发现β-LG 由乳腺合成并分泌到乳中（Madureira et al.，2007）。

它在水中溶解度较低（Divsalar et al.，2009），分子质量为 18.20～18.40 kDa，共含有 162 个氨基酸残基（图 3.1）（Yadav et al.，2015）。β-LG 的粒径约为 2 nm，等电点约为 5.2（Bolder et al.，2007）。

Leu(1)-Ile-Val-Thr-Gln-Thr-Met-Lys-Gly-Leu-Asp-Ile-Gln-Lys-Val-Ala-Gly-Thr-Trp-Tyr-Ser-Leu-Ala-Met-
Ala-Ala-Ser-Asp-Ile-Ser-Leu-Leu-Asp-Ala-Gln-Ser-Ala-Pro-Leu-Arg(Se)-Val-Tyr-Val-Glu-Glu-Leu-Lys-
Pro-Thr-Pro-Glu-Gly-Asp-Leu-Glu-Ile-Leu-Leu-Gln-Lys-Trp-Glu-Asn-Asp-Glu-Cys-Ala-Gln-Lys-Lys-Ile-Ile-
Ala-Glu-Lys-Thr-Lys-Ile-Pro-Ala-Val-Phe-Lys-Ile-Asp-Ala-Leu-Asn-Glu-Asn-Lys-Val-Leu-Val-Leu-Val-
Leu-Asp-Thr-Asp-Tyr-Lys-Lys-Tyr-Leu-Leu-Phe-Cys-Met-Glu-Asn-Ser-Ala-Glu-Pro-Gln-Ser-Leu-Val-Cys-
Gln-Cys-Leu-Val-Arg(Se)-Thr-Pro-Glu-Val-Asp-Asp-Glu-Ala-Leu-Glu-Lys-Phe-Asp-Lys-Ala-Leu-Lys-Ala-
Leu-Pro-Met-His-Arg(Se)-Leu-Ser-Phe-Asn-Pro-Thr-Gln-Leu-Glu-Glu-Gln-Cys-His-Ile(162)

图 3.1　牛乳 β-LG A 型变体序列

目前已确认七种不同的 β-LG 遗传变体，但是工业生产的 β-LG 中 A 和 B 两种变体是最常见的。β-LG 的 A 和 B 两种变体与牛乳的形成和组成有关，因此不同品种的牛乳之间会存在差异（Mezanieto et al.，2012）。图 3.1 给出了 A 型 β-LG 的序列。B 型与 A 型两种变体在氨基酸序列的第 64 位（A 型 Asp→B 型 Gly）和第 118 位（A 型 Val→B 型 Ala）存在差异（Divsalar et al.，2009）。两种变体在热稳定性、电位、某些基团的反应性，以及在聚合速率、聚合物的粒径及随后形成凝胶的强度等热诱导聚合行为上也存在差异（Schokker et al.，2000）。

β-LG 是一种由 β 折叠主导的蛋白质（Liu et al.，2006）。图 3.2 为 β-LG 晶体单个亚基的带状图。A～I 是结构中的九条链，无规则卷曲被作出了标记。A～D 和 E～H 分别与一个 β 折叠和末端 COOH 形成一个平面。A～H 反平行的 β 折叠扭曲成一个锥形桶，带有一个保护性的 α 螺旋，而 E～F 则是桶的出入口。该出入口

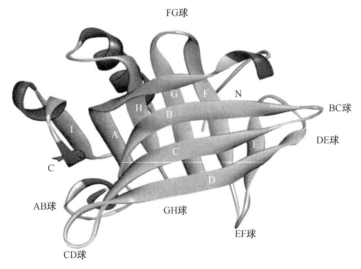

图 3.2　牛乳 β-LG 晶体单个亚基结构
该结构（3uew）来自结构生物信息学研究合作实验室（RCSB）蛋白质数据库（http://www.rcsb.org）

在酸性条件下闭合，在碱性条件下打开，所以一些疏水基团可以与 A 链和 H 链的侧链残基结合。因此，牛乳β-LG 被归类为脂质运载蛋白（Adams et al., 2006），这类蛋白具有运输营养成分、参与细胞调控、运输信息素、隐蔽色和分泌前列腺素酶等功能。大多数的脂质运载蛋白能够在一个内部的结合口袋与疏水小分子结合（Brownlow et al., 1997）。β-LG 能够与不同的疏水配体如脂肪酸、血晶素、玫瑰树碱、芳香烃、致癌碳氢化合物等发生剧烈反应（Divsalar et al., 2009）。据文献记载，β-LG 的单体至少含有两个疏水性结合位点，一个位于内部，另一个位于表面，介于β桶和α螺旋之间（Paul et al., 2014）。这种特性使β-LG 和乳清蛋白质成为极好的纳米颗粒载体和微胶囊壁材。

对于婴儿来说，β-LG 是乳清组分的主要变应原，存在于牛乳中，但在人乳中不存在（Yong et al., 2010）。它甚至可以在极低的浓度诱发过敏反应（通常低于μmol/L 水平）。β-LG 由于其特殊的结构能够耐受胃蛋白酶的水解作用（Reddy et al., 1988）。β-LG 的疏水基团和芳香族氨基酸侧链等潜在的酶切位点大多被包埋在β-LG 的核心部位，因此可以避免与酶接触（Hernández-Ledesma et al., 2006）。结构的改变可以增加β-LG 对蛋白质水解作用的易感性。β-LG 溶液在 90~100℃热处理 5~10 min 引起的结构或者构象改变能够促进其与胃蛋白酶接触并且增加胰蛋白酶的水解度（Guo et al., 1995）。当处理温度增加到 60~80℃时，原本大部分被包埋在β-LG 内部能够被嗜热菌蛋白酶识别的切割位点能够暴露出来（Hernández-Ledesma et al., 2006）。

未折叠的β-LG 以残留的β-结构为表征，该结构通过二硫键保持稳定（Kuwajima et al., 1996）。每一种β-LG 都在第 66、106、119、121 和 160 点位含有 5 个半胱氨酸（图 3.1）。4 个半胱氨酸残基形成 2 个二硫键（Cys66-Cys160、Cys106-Cys119），还有 1 个游离的半胱氨酸（121）（Guo and Wang, 2016；Walsh, 2014）。121 位半胱氨酸的游离—SH 是活性基团，容易发生氧化反应。SH/SS 交互作用和/或 SH/SH 氧化反应是以β-LG 为主要成分的乳清蛋白质发生聚合和形成凝胶的主要原因。乳清蛋白质的聚合和成胶性因其在食品领域不同体系的广泛应用而成为其主要的功能特性。

β-LG 的四级结构是由体系的 pH 和温度共同决定的。室温条件下，在 pH 5.2~7，包括天然牛乳的 pH，β-LG 作为二聚体存在于溶液中（Chatterton et al., 2006）。在中性条件下，二聚体为一个长 6.95 nm、宽 3.6 nm 的椭圆形，当加热到约 70℃时，二聚体分解成单体并且部分展开（Schokker et al., 1999）。当 pH 介于 3.5~5.2 时，β-LG 可形成一种分子质量约为 140000 kDa 的八聚体（Madureira et al., 2007）。当 pH 高于 8.0 或者低于 3.0 时，β-LG 能够分解成单体（Ng et al., 2016）。

3.1.2 β-乳球蛋白的分离与制备

β-LG 在牛乳中的平均浓度约为 3.2 mg/mL（Pan et al.，2007），其他主要成分的浓度分别为：α-LA 约为 0.1 mg/mL，BSA 约为 0.04 mg/mL，Ig 约为 0.08 mg/mL（Adams et al.，2006）。到目前为止，β-LG 已经从牛乳、干酪乳清或乳清蛋白质粉，包括 WPC 和 WPI 中实现了分离。当对牛乳中 β-LG 进行分离/纯化时，占总蛋白 80%（w/w[①]）的酪蛋白可以通过添加有机或无机酸等电点沉降法被去除（Madureira et al.，2007）。在 pH 为 4.6 时，乳清蛋白质为溶液状态，而酪蛋白变成沉淀。酪蛋白也可以通过凝乳酶凝固法去除，该法使乳清蛋白质以副产物的形式释放出来（Madureira et al.，2007）。酪蛋白还可以通过超速离心的方式去除（Jongh et al.，2001）。去除酪蛋白获得的乳清含有 200 多种不同的复合物（Michaela et al.，2008），包括 6～8 g/L 蛋白质、44～52 g/L 乳糖和 2.5～7.2 g/L 矿物质（Guo and Wang，2016）。乳清蛋白质粉通常经过超滤或者超滤/渗滤结合的方法（UF/DF）进行脱盐，再经浓缩和喷雾干燥获得（Rabiey and Britten，2009）。提取有价值的蛋白质主要有三种方法，即层析法（如离子交换法和疏水吸附法）、膜分离法（如传统的压力驱动法和电分离法）或者二者组合的方法（El-Sayed and Chase，2011）。基于 β-LG 的化学性质，不同的方法如超滤、离子交换色谱、沉淀法、试剂络合法或者几种方法相结合被用于 β-LG 的纯化（Montilla et al.，2007）。

1. 超滤和离子交换色谱法

以乳清为原料分离纯化 β-LG 就是要去除乳清中所有其他组分。乳清首先经过离心和微滤处理去除主要的固形物如胶体物质、悬浮的酪蛋白颗粒和脂质，然后采用渗滤法去除大部分乳糖、矿物质和盐来富集乳清蛋白质。微滤和渗滤后得到的滤液是各种乳清蛋白质的混合物。超滤作为一种在乳品工业中公认的良好的加工操作已被用于分离 β-LG。它包括各种不同的膜过滤，其中产生的压力或者浓度梯度等能够促使物料通过半透膜完成分离。截留分子质量为 10～30 kDa 的滤膜通常被用来分离分子质量约为 18 kDa 的 β-LG。

为了提高超滤后 β-LG 的纯度，通常会进行离子交换层析处理。离子交换层析后再用 10～30 kDa 的滤膜进行超滤处理，从生产酪蛋白的乳清中纯化 β-LG，最终产物的纯度可以达到 87.6%（Bhattacharjee et al.，2006）。β-LG B 和 β-LG A 能够很容易被强阴离子交换剂季氨乙基化合物吸附并用 0.05 mol/L tris-HCl 缓冲液（pH 6.8）制备的线性浓度梯度（0.1～0.25 mol/L）的 NaCl 溶液洗脱。β-LG B 和 β-LG A 的产率为 3570 mg/L 乳清（Ye et al.，2000）。有一种阴离子交换色谱法被用来从 WPC80 中分离 β-LG。该方法通过离子柱和盐溶液梯度洗脱来增加洗脱缓冲液（20 mmol/L tris-HCl 加上 0～1 mol/L NaCl）的离子强度，β-LG 的两种变体

---

① w/w 表示质量分数。

采用这种方法分离的回收率可达 60.5%（Santos et al., 2012）。

由于加工成本昂贵，废水体积大，离子交换色谱无法应用于商业生产。对于等电点有一定的差异，并且是膜分子量截留值的 1/20～1/15 的蛋白质，可以用带电超滤膜进行分离。与无电膜相比，α-LA 和 β-LG 的分馏选择性有可能提高 180%，这样就可以获得纯度分别为 87%和 83%的α-LA 和β-LG（Arunkumar and Etzel, 2014）。

2. 沉淀法

尽管超滤是一种高选择性的有效方法，但是它耗时较多且对条件要求苛刻。在一定条件下使β-LG 发生沉淀的方法可以用来替代超滤法。有人提出了一种改良的β-LG 分离方法。第一步，在低 pH 状态或者降低 pH，向乳清中添加硫酸铵，在加热或不加热的条件下促使除β-LG 以外的所有蛋白质发生沉淀来获得β-LG（Schlatterer et al., 2004）；第二步，利用弱酸性的聚合树脂进行阳离子交换色谱层析（Lozano et al., 2008）。

β-LG 是乳清中最不易被三氯乙酸引起沉淀的组分。当牛乳中的酪蛋白通过酸沉法被去掉后，向乳清层添加含量为 3%的三氯乙酸。除了β-LG 以外的所有乳清蛋白质均会形成沉淀，然后过滤可获得β-LG（Fox et al., 1967）。然而，由于蛋白质的原始结构是由 pH、温度、压力和溶剂等共同决定的，所以通过沉淀法所得蛋白质的功能性有可能会随着结构的改变而发生变化。

3. 胃蛋白酶水解法

由于β-LG 对胃蛋白酶的抗性，因此可以通过胃蛋白酶水解和膜分离的方法提取浓缩乳清蛋白质中的β-LG（Kinekawa and Kitabatake, 1996）。乳清蛋白质首先被猪胃蛋白酶水解，然后被硫酸铵沉淀，再通过透析膜（分子质量 20 kDa）或者超滤膜（分子质量 30 kDa）过滤来获得纯化的β-LG。用该法制备的β-LG 与标准品没有差异。

4. 试剂络合法

牛乳β-LG 也可以通过与试剂如壳聚糖络合的方法进行纯化。酸乳清中的β-LG 可以通过静电作用与壳聚糖形成β-LG-壳聚糖络合物再经离心进行分离。壳聚糖可以通过酸性水解被去除。获得的蛋白质可以保持高度的原始状态并且纯度高达 95%（Montilla et al., 2007）。

3.1.3　β-乳球蛋白的生物特性

在生理条件下，牛乳β-LG 以二聚体的形式存在，并与一些疏水小分子和两性化合物（如磷脂）的运输有关（Liu et al., 2006）。它也是维生素 A 的结合蛋白（Yang et al., 2010）。此外，β-LG 富含半胱氨酸，而半胱氨酸是刺激谷胱甘肽合成的必需氨基酸（Mcintosh et al., 1995）。现已对有关β-LG 的抗菌、血管紧张素转换酶（ACE）抑制和抗氧化活性等进行了体外研究。

1. 抗微生物活性

研究表明，正电荷和疏水性可能是抗菌活性的重要因素（Pellegrini，2003）。β-LG 的抗菌活性包括抗菌和抗病毒作用。β-LG 能够抑制金黄色葡萄球菌和乳房链球菌生长（Chaneton et al.，2011），但对大肠杆菌、枯草芽孢杆菌、无害李斯特菌和变形链球菌没有任何抗菌活性（Chevalier et al.，2001）。与完整的β-LG 相比，β-LG 衍生物具有较强的抗菌活性。β-LG 的肽类 VAGTWY f（15~20）、AASDISLLDAQSAPLR f（25~40）、IPAVFK f（78~83）和 VLVLDTDYK f（92~100）通过胰蛋白酶水解进行分离和表征。据报道，这些片段可以有效抑制革兰氏阳性菌（Pellegrini et al.，2001）。

β-LG 的化学修饰可能会影响该蛋白质的抗菌性能。酰胺化的β-LG 对荧光假单胞菌、莓实假单胞菌和枯草芽孢杆菌的细胞有很强的杀菌作用，但对大肠杆菌、粪肠球菌、伤寒沙门氏菌和单核细胞增生李斯特菌的杀菌作用弱得多。无论是天然的还是酰胺化的β-LG，都不能有效地拮抗酿酒酵母和青霉菌（Pan et al.，2007）。β-LG 与阿拉伯糖、半乳糖、葡萄糖、乳糖、鼠李糖和核糖发生糖基化后对大肠杆菌、枯草芽孢杆菌、无害李斯特菌和变形链球菌无抗菌作用（Chevalier et al.，2001）。

β-LG 分子中加入 3-羟基苯酐（3-HP）后对单纯疱疹病毒 1（HSV-1）具有抗病毒活性。同时，通过胰蛋白酶、糜蛋白酶和胃蛋白酶对β-LG 进行水解可产生几个具有抗热活性的肽片段（Oevermann et al.，2003）。

2. 血管紧张素转换酶（ACE）抑制活性

完整的β-LG 具有很差的 ACE 抑制活性，而对该蛋白进行消化后可产生较高的 ACE 抑制指数。如上所述，β-LG 是抗胃蛋白酶水解的，因为其结构紧密并且酶切割位点位于β-LG 核心。各种方法被用于改善水解以产生活性肽。ACE 抑制肽（casokinins）于 1980 年首次从α-酪蛋白和β-酪蛋白中得到分离（Maruyama and Suzuki，1982），然后乳激肽（lactokinins）在 1990 年从β-LG 中得到纯化（Mullally et al.，1996）。表 3.2 总结了采用不同酶在β-LG 中发现的 ACE 抑制肽。结果表明，与消化酶相比，通过微生物酶从β-LG 中获得的肽具有更高的活性（Jelen，2010）。β-LG 与嗜热菌蛋白酶在 60℃ 和 80℃ 处理 20~30 min 产生的肽 LQKW f（58~61）是强效 ACE 抑制剂（IC_{50} 值 34.7 μmol/L）（Hernández-Ledesma et al.，2006）。

表 3.2　β-LG 中 ACE 抑制肽的初级序列总结

序列	IC_{50} 值	酶	参考文献
MKG f(7~9)	71.8 μmol/L	嗜热菌蛋白酶	Hernández-Ledesma et al.，2016
RL f(148~149)	2439 μmol/L	嗜热菌蛋白酶	Hernández-Ledesma et al.，2016
IRL f(147~149)		嗜热菌蛋白酶	Hernández-Ledesma et al.，2016
VFK f(81~83)	1029 μmol/L	嗜热菌蛋白酶	Hernández-Ledesma et al.，2016

续表

序列	IC$_{50}$值	酶	参考文献
LDIQK f(10～14)	27.6 μmol/L	嗜热菌蛋白酶	Hernández-Ledesma et al., 2016
VAGTWY f(15～20)	1682 μmol/L	嗜热菌蛋白酶	Hernández-Ledesma et al., 2016
LRVY f(39～42)	205.6 μmol/L	嗜热菌蛋白酶	Hernández-Ledesma et al., 2016
SAPLRVY f(36～42)	205.6 μmol/L	嗜热菌蛋白酶	Hernández-Ledesma et al., 2016
LQKW f(58～61)	34.7 μmol/L	嗜热菌蛋白酶	Hernández-Ledesma et al., 2016
Ala-Leu-Pro-Met-His-Ile-Arg f (142～148)	42.6 μmol/L	胰蛋白酶	Implvo et al., 2007；Mullally et al., 1996
Gly-Leu-Asp-Ile-Gln-Lys f(9～14)	580 μmol/L	胃蛋白酶和胰蛋白酶	Pihlanto-Leppälä et al., 1998
Val-Ala-Gly-Thr-Trp-Tyr f(15～20)	1682 μmol/L	胃蛋白酶和胰蛋白酶	Pihlanto-Leppälä et al., 1998

3. 抗氧化活性

β-LG 在牛乳中起着关键的抗氧化作用。β-LG 的抗氧化活性包括通过氨基酸残基清除自由基和螯合促氧化过渡态金属离子。它是一种温和的抗氧化剂，其效力低于维生素 E。Cys121 在β-LG 的抗氧化性质中起着至关重要的作用，通过加热交联游离巯基或通过羧甲基化修饰β-LG 阻断巯基，能够导致其丧失大量抗氧化活性（Liu et al., 2007）。在具有抗氧化能力的氨基酸残基中，半胱氨酸的抗氧化能力最强，因为它与脂质衍生的自由基和 Cu^{2+} 催化衍生的自由基有很强的相互作用（Sharp et al., 2004）。油包水乳剂中的β-LG 对延缓脂质氧化反应十分有效。研究表明，β-LG 中的半胱氨酸能够在色氨酸之前被氧化，并且检测结果显示这两种氨基酸残基的氧化反应均优先于脂质的氧化反应（Elias et al., 2005）。

酶解后的β-LG 比完整的β-LG 具有更高的抗氧化活性。在水包油的乳液连续相（pH 7）中，β-LG 的酶解物比同等浓度的β-LG 能更有效地延迟脂质的氧化反应（Elias et al., 2006）。与β-LG 相比，其酶解产物具有较高的过氧自由基清除能力和铁结合能力。通过对三种氧化不稳定的氨基酸残基（Tyr、Met 和 Phe）进行氧化速率测试，发现 Tyr 和 Met 残基能够清除自由基，具有提高食品脂质分散体系氧化稳定性的潜力（Elias et al., 2006）。

3.2　α-乳白蛋白

3.2.1　α-乳白蛋白的化学特性

α-乳白蛋白（α-LA）是一种小的致密球蛋白，分子质量约为 14 kDa，等电点范围为 4.2~4.5。它是一种强 Ca^{2+}结合蛋白。图 3.3 为α-LA 的初级结构。它的单条肽链由 123 个氨基酸组成（Konrad and Kleinschmidt，2008）。人乳和牛乳最重要的区别之一是蛋白质组成不同（Heine et al.，1991）。人乳和牛乳的α-LA 中氨基酸的百分比存在差异（表 3.3）。

Glu(1)-Gln-Leu-Thr-Lys-Csy-Glu-Val-Phe-Gln-Glu-Leu-Lys-Asp-Leu-Lys-Gly-Tyr-Gly-Gly-Val-Ser-Leu-Pro-Glu-Trp-Val-Cys-Thr-Thr-Phc-His-Thr-Ser-Gly-Tyr-Asp-Thr-Glu-Ala-Ile-Val-Glu-Asn-Asn-Gln-Ser-Thr-Asp-Tyr-Gly-Leu-Phe-Gln-Ile-Asn-Asn-Lys-Ile-Trp-Cys-Lys-Asn-Asp-Gln-Asp-Pro-His-Ser-Ser-Asn-Ile-Cys-Asn-Ile-Ser-Cys-Asp-Lys-Thr-Leu-Asn-Asn-Asp-Leu-Thr-Asn-Asn-Ile-Met-Cys-Val-Lys-Lys-Ile-Leu-Asp-Lys-Val-Gly-Ile-Asn-Tyr-Trp-Leu-Ala-His-Lys-Ala-Leu-Cys-Ser-Glu-Lys-Leu-Asp-Gln-Trp-Leu-Cys-Glu-Lys-Leu-OH(123)

图 3.3　牛乳α-LA 的初级结构

资料来自 Stănciuc and Râpeanu（2010）

表 3.3　人乳和牛乳α-LA 氨基酸百分比的比较分析　　（单位：%）

氨基酸	人乳	牛乳
色氨酸	1.8	5.9
苯丙氨酸	4.4	4.0
亮氨酸	10.1	10.5
异亮氨酸	5.8	6.1
苏氨酸	4.6	5.0
甲硫氨酸	1.8	0.9
赖氨酸	6.2	10.3
缬氨酸	6.0	4.3
组氨酸	2.3	2.6
精氨酸	4.0	1.0
胱氨酸	1.7	5.3
脯氨酸	8.6	1.4
丙氨酸	4.0	1.9
天冬氨酸	8.3	16.8
丝氨酸	5.1	4.5
谷氨酸	17.8	11.5
甘氨酸	2.6	3.3
酪氨酸	4.7	4.7

α-LA 通过 4 个二硫键维持稳定（Cys6-Cys120、Cys61-Cys77、Cys73-Cys91 和 Cys28-Cys111），并且不含游离巯基（Schokker et al.，2000）。图 3.4 为α-LA 的

三级结构，它被一个裂缝分为两个域，一个主要是螺旋结构（α域），另一个β折叠含量显著（残基35~85，β域）（Redfield et al.，1999）。这两个结构域通过钙结合环连接。α域由3个pH稳定的α螺旋（残基5~11、23~24和86~98）、1个pH依赖的α螺旋（H4-105~110）和2个短的3_{10}螺旋（每个螺旋包括3个残基和1个含有10个原子的内键环，残基18~20和115~118）组成。较小的β域由一系列环状结构、1个小的三股反平行的L折叠片（残基41~44、47~50和55~56）和1个短的3_{10}螺旋（残基77~80）组成（Permyakov and Berliner，2000）。这种蛋白质有一个很强的钙结合位点，它是由3个Asp残基（82、87和88）的羧基和肽骨架的2个羰基（79和84）在2个螺旋之间形成的氧配体（Permyakov and Berliner，2000）。牛乳中的α-LA以1:1的物质的量比与钙紧密结合。据报道，α-LA也可以通过N端的结合位点与Zn结合。它还可以与Mn^{2+}、Cd^{2+}、Mg^{2+}、Co^{2+}和Hg^{2+}结合。

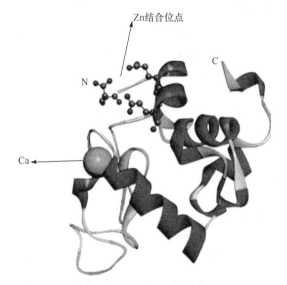

图3.4 α-LA的三级结构

该结构（1 f6s）来自结构生物信息学研究合作实验室（RCSB）蛋白质数据库（http://www.rcsb.org）

熔融的球状体是球状蛋白质在天然状态和完全展开状态之间的中间体（Arai and Kuwajima，1996）。α-LA有几种部分折叠的中间态，这是许多人正在研究的焦点。在酸性pH和高温的apo状态下，α-LA为经典的熔融球状（Permyakov and Berliner，2000）。

3.2.2 α-乳白蛋白的分离

α-LA是人乳中的一种主要蛋白质。向婴儿配方奶粉中加入α-LA能够使其组成更加接近人乳。α-LA 的分离受到越来越多的关注。通常利用等电点沉淀法（Bramaud et al., 1997）和离子交换层析法（Gerberding and Byers, 1998）等方法来分离α-LA。

调节酸或甜乳清的 pH 使其接近α-LA 的等电点而发生沉淀的分离方法较复杂，最优的 pH 为 3.9，此时获得的α-LA 纯度最高。离子交换色谱法往往可以比等电点沉淀法获得更高纯度的α-LA。调节 pH 后，利用 50%硫酸铵将牛乳乳清中的α-LA 组分解析出来，然后通过阴离子交换层析法用二乙氨基乙基琼脂糖快速流动纯化。α-LA 纯度约为 85%，交叉反应性仍为 93.2%（Mao et al., 2016）。

3.2.3 α-乳白蛋白的功能特性

在新生儿时期，人体内的α-LA 变体具有多种生理功能。在乳腺中，它能够参与乳糖合成，促进乳汁的产生和分泌（Stănciuc and Râpeanu, 2010）。图 3.5 显示了人类乳糖的生物合成。乳糖合成涉及三个反应，在第三步，乳糖合成酶通过将半乳糖基转移到葡萄糖而生成乳糖。乳糖合成酶由两个亚基组成，一个是β-1,4-半乳糖基转移酶，另一个是α-LA。当复合物形成时，酶对葡萄糖的亲和力可提高 1000 倍（Stănciuc and Râpeanu, 2010）。在哺乳动物中，α-LA 含量与乳糖含量呈正相关。

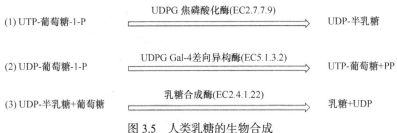

图 3.5 人类乳糖的生物合成

UTP：尿苷三磷酸；UDP：尿苷二磷酸；UDPG：尿苷二磷酸葡糖

α-LA 的优质营养价值（蛋白效率比为 4.0，乳清效率比为 3.6，酪蛋白效率比为 2.9）和良好的生物学价值被高度认可（Lucena et al., 2006）。虽然牛乳在许多方面与人乳不同，但它是最常用于婴幼儿配方奶粉制造的营养来源。α-LA 因其色氨酸含量高，可添加到婴幼儿配方奶粉中（Arunkumar and Etzel, 2014）。

用胰蛋白酶对α-LA 进行消化处理可得到两个片段：EQLTK（残基 1～5）和 GYGGVSLPEWVCTTF ALCSEK[残基（17～31）S—S（109～114）]。这两条多肽链通过二硫键结合，并且能够与 CKDDQNPH ISCDKF 的糜蛋白酶水解物[残基（61～68）S—S（75～80）]通过二硫键结合在一起发挥抗菌活性，该多肽对革兰氏阳性菌的抑制作用最强（Pellegrini et al., 1999）。

3.3 牛血清白蛋白

牛血清白蛋白（BSA）作为血浆蛋白的主要成分之一是单链球状的非糖蛋白（Liu et al.，2004）。BSA 分子质量为 66 kDa（Sklar et al.，1977）。BSA 的一级结构与人血清白蛋白（HSA）非常相似（表 3.4）。BSA 由 583 个氨基酸残基组成，而 HSA 为 585 个。BSA 和 HSA 均含有 35 个半胱氨酸残基，形成 17 个二硫键，并留下 1 个游离的巯基。BSA 在 134 和 213 上含有 2 个色氨酸残基以及 20 个酪氨酸残基。

表 3.4　人血清白蛋白与牛血清白蛋白的一级结构组成　　　（单位：个）

	Ala	Cys	Asp	Glu	Phe	Gly	His	Ile	Lys	Leu
HSA	62	35	36	62	31	12	16	8	59	61
BSA	47	35	40	59	27	16	17	14	59	61
	Met	Asn	Pro	Gln	Arg	Ser	Thr	Val	Trp	Tyr
HSA	6	17	24	20	24	24	28	41	1	18
BSA	4	14	28	20	23	28	33	36	2	20

图 3.6 显示了 BSA 的结构，BSA 是由 α 螺旋（67%）主导的，其余为依次排

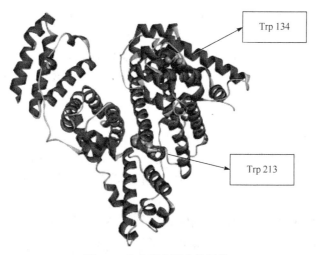

图 3.6　牛血清白蛋白的结构

该结构（3v03）来自结构生物信息学研究合作实验室（RCSB）蛋白质数据库（http://www.rcsb.org）

列和扩展的多肽或没有β折叠的灵活区域（Carter and Ho，1994）。它是一种心形蛋白质，有三个同源结构域（Ⅰ、Ⅱ、Ⅲ）。这17个二硫键将结构域划分为9个环（L1～L9）（Papadopoulou et al.，2005）。每个域有两个子域（如ⅠA和ⅠB）。BSA有两个色氨酸残基（Trp134和Trp213），分别位于ⅠA和ⅡA亚区（Kandagal et al.，2006）。Trp213位于蛋白质的疏水结合口袋内，参与表面疏水腔的形成，在药物的转运载体中起着关键作用。

血清白蛋白作为循环系统的主要可溶性蛋白成分，具有多种生理功能。它们有助于维持血液的胶体渗透压，并负责维持血液的pH（Klajnert and Bryszewska，2002）。白蛋白最突出的特性是它能够可逆地结合多种配体（Carter and Ho，1994）。该蛋白质的主要生理作用是血清中游离脂肪酸的载体。研究结果表明，在10^6～10^8 mol/L的范围内，每摩尔乳清白蛋白可以结合5～6 mol的长链脂肪酸（Sklar et al.，1977）。

3.4 乳铁蛋白

1939年，乳铁蛋白（Lf）首次从牛乳中作为一种未知的"红色组分"被分离出来（Sørensen and Sørensen，1939）。它存在于牛乳中，少数存在于外分泌液体如胆汁和眼泪中。Lf是转铁蛋白家族中分子质量为80 kDa的铁结合糖蛋白。它在多肽链中有2个碳水化合物侧链（约7.2%），含有15～16个甘露糖，5～6个半乳糖，10～11个N-乙酰半乳糖胺，1个单位唾液酸和1个单位岩藻糖。Lf属于等电点约为8.7的碱性蛋白质，具有调节粒细胞生成、激活NK细胞、抗宫颈癌病毒等多种生物活性（Wassef et al.，2008）。Lf对乳腺预防乳腺炎起着重要的作用。它对多种细菌和真菌具有天然的防御功能（Hwang et al.，1998）。Lf能够有效对抗大肠杆菌（Chaneton et al.，2011）。最初认为，Lf的抗菌活性完全归因于其固铁能力。Lf被认为在肠黏膜对铁的吸收发挥作用，还可以作为一种抑菌剂，从需要铁的细菌中截留铁。Lf对蛋白质水解具有相对抗性（Lönnerdal and Iyer，1995）。然而，最近发现Lf在N端附近的肽段比完整的Lf具有更强的杀菌作用，该肽被称为乳铁蛋白肽（Lfcin），对微生物具有广泛的杀伤力（Hwang et al.，1998）。

牛乳铁蛋白和人乳铁蛋白分别由689个和691个氨基酸残基组成；序列一致性程度达69%（Permyakov and Berliner，2000）。表3.5表明了牛乳铁蛋白和人乳铁蛋白中氨基酸残基数量的差异（Pierce et al.，1991）。

表 3.5　牛乳铁蛋白和人乳铁蛋白中氨基酸残基的数量

	牛乳铁蛋白	人乳铁蛋白
丙氨酸	67	63
脯氨酸	30	35
精氨酸	39	43
赖氨酸	54	46
天冬氨酸	29	33
缬氨酸	47	48
色氨酸	13	10
半胱氨酸	34	32
苏氨酸	36	31
异亮氨酸	15	16
丝氨酸	45	50
谷氨酸	40	42
苯丙氨酸	27	30
甲硫氨酸	4	5
亮氨酸	65	58
甘氨酸	48	54
酪氨酸	22	21
天冬氨酸	36	38
组氨酸	9	9
残基总数	689	691

注：资料来自 Pierce et al. (1991)。

图 3.7 为牛乳中 Lf 的结构。它由两个同源裂片（N 端和 C 端半部）组成。每个裂片分为两个子域,分别是 N1(残基 1~90 和 251~333)和 N2(残基 91~250)、C1(残基 345~431 和 593~676)和 C2(残基 432~592)。域间裂缝的深处有一个铁结合位点。其余的残基 334~344 序列代表了中枢,这是一个螺旋构象,有 3 个弯曲,在结构域打开和关闭时发挥作用(Permyakov and Berliner, 2000)。Holo-Lf [图 3.7（a）]和 Apo-Lf[图 3.7（b）]在 C1 和 N1 域具有完全相同的结构,但在 C2 和 N2 域中存在差异。Holo-Lf 具有紧密结合的 C2 和 N2 结构域,而 Apo-Lf 是开放结构。因此,Holo-Lf 比 Apo-Lf 更稳定。

牛乳中 Lf 的含量从初乳的 7 g/L 到熟乳的 1 g/L 不等。为了分离 Lf,由于回收率低、乳清需求量过大和分离过程的复杂性,更有效的方法取代了传统的柱基色谱法。一种用合成的微米级单分散超顺磁性聚甲基丙烯酸缩水甘油酯（PGMA）颗粒与肝素偶联，以氯化钠溶液为洗脱剂，从牛乳清中纯化 Lf 的方法被提出。该方法是一步纯化，得到的 Lf 纯度高于商业标准蛋白（Chen et al., 2007）。

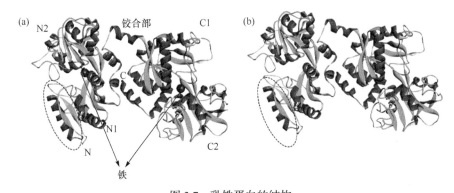

图 3.7　乳铁蛋白的结构
（a）含饱和铁的乳铁蛋白；（b）不含铁乳铁蛋白。
该结构（1 biy）来自结构生物信息学研究合作实验室（RCSB）蛋白质数据库（http://www.rcsb.org）

3.5　免疫球蛋白

牛免疫球蛋白（Ig）是初乳中免疫活性物质的重要组成部分。它属于球状蛋白质家族，具有一系列生物保护活性。牛初乳免疫球蛋白的生物学功能是为新生小牛提供足够的被动免疫保护，以防止微生物感染（Mehra et al., 2006）。免疫球蛋白被选择性地从血清中转运到乳腺，因此初乳中含有非常高浓度的免疫球蛋白（40～200 mg/mL）（Korhonen et al., 2000）。一些研究表明，从高免疫奶牛的初乳中摄取的免疫球蛋白可以预防轮状病毒（Tacket et al., 1988；Fukumoto et al., 1994）、肠毒性大肠杆菌（Mietens et al., 1979；Tacket et al., 1988）、福氏志贺氏菌（Tacket et al., 1992）和隐孢子虫（Tzipori et al., 1986；Soave and Armstrong, 1987）对婴儿和成人造成感染（Li et al., 2010）。

免疫球蛋白包括 IgM、IgA、IgG、IgE 和 IgD 等几类。所有单体免疫球蛋白具有相同的基本分子结构，由两条重链和两条轻链组成。图 3.8 给出了免疫球蛋白的结构示意图。重链和轻链通过二硫键连接在一起，使免疫球蛋白分子呈经典的 Y 形。重链和轻链都有恒定区域（A）和可变区域（B）（Considerations, 2006）。

IgG、IgM 和 IgA 是乳腺分泌的主要免疫球蛋白（Hurley and Theil, 2011）。IgG 以单体形式出现（约 160 kDa），IgM（约 1000 kDa）和 IgA（约 370 kDa）形成聚合 Ig。牛 IgG 在初乳中含量约为 50 mg/mL。免疫球蛋白 IgG 可分为 IgG1 和 IgG2 两个亚型（Bogahawaththa et al., 2017）。IgG1 是最丰富的 Ig 同型，约占 Ig

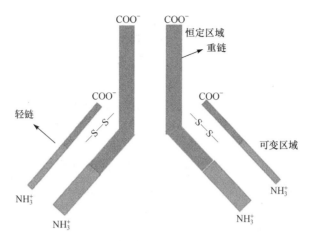

图 3.8 免疫球蛋白结构示意图

总含量的 80%（Chen et al., 2010）。IgG 具有较高的等电点（IgG1 的等电点为 5.5～6.8，IgG2 的等电点为 7.7～8.3），分子质量大于 150 kDa，可用聚苯乙烯阴离子交换器 IRA93 和 Amicon YM100 膜从 HCl-酪蛋白和结肠乳清中获得（Xu et al., 2000）。牛 IgM 存在于血清、初乳和牛乳中。IgM 在原发性免疫反应、补体结合和血清凝集抗体中起重要作用。IgM 似乎与非支原体的寄生虫感染密切相关。牛 IgA 在牛乳和初乳中表现为"分泌型 IgA"（Butler, 1969）。

3.6 微量蛋白质

乳清蛋白质还含有一些数量少但活性高的蛋白质，如生长因子、乳过氧化物酶、乳脂肪球膜蛋白和维生素结合蛋白。

3.6.1 生长因子

生长因子是能与不同类型细胞结合的小蛋白，包括胰岛素样生长因子-Ⅰ（IGF-Ⅰ）和转化生长因子-β2（TGF-β2），其含量分别为 5～100 mg/mL 和 10～70 mg/mL（Wassef et al., 2008）。生长因子能够促进新生小牛生长发育（Pakkamen and Aalto, 1997）。IGFs 为单链多肽，分子质量约为 7.6 kDa。每个 IGF 分子含有 3 个二硫键（Pakkamen and Aalto, 1997）。

3.6.2 乳过氧化物酶

乳过氧化物酶（LP）是过氧化物家族的成员。乳过氧化物酶的主要功能是利

用 H_2O_2 对分子进行氧化。牛的乳过氧化物酶由单个多肽链构成，有 612 个氨基酸残基，分子质量约为 78 kDa。它是一种等电点较高（9.6）的碱性蛋白质（Kussendrager and van Hooijdonk，2000）。乳过氧化物酶的含量约为 30 mg/L，据泌乳情况而变化（Björck et al.，1975）。它是牛乳中的一种天然抗菌药物。催化剂（LP）和两种反应物（H_2O_2 和 SCN^-）构成了天然的乳过氧化物酶抗菌体系，已被证实对各种革兰氏阳性和革兰氏阴性微生物具有抑菌和杀菌效果（Wolfson and Sumner，1993）。乳过氧化物酶催化 H_2O_2 氧化硫氰酸盐（SCN^-），能够生成具有抗菌性能的中间产物，对保护泌乳乳腺和新生儿肠道免受病原微生物侵害具有重要作用（Seifu et al.，2005）。乳过氧化物酶能够阻止嗜冷细菌的增殖，延长原料乳在低温条件下的储藏期（Björck，1978）。牛乳过氧化物酶具有相对耐热性，该酶在74℃时仅可被短时间巴氏杀菌部分灭活，可保留足够活性来催化硫氰酸盐与 H_2O_2 的反应（Wolfson and Sumner，1993）。

3.6.3 乳脂肪球膜蛋白

天然的牛乳脂肪球表面有一层保护膜，通常被称为牛乳脂肪球膜（MFGM）（Ye et al.，2002）。MFGM 横截面约为 10~20 nm 厚，具有乳化剂的功效，可以保护脂肪球不发生聚合和酶解（Elloly，2011）。膜本身是由蛋白质、糖蛋白、酶、中性脂质和磷脂等极性脂质组成的复杂混合物（Fong et al.，2007）。其中 MFGM 蛋白占膜成分的 25%~60%，占牛乳总蛋白含量的 1%~2%。虽然数量相对较少，但 MFGM 蛋白具有较高的生物活性，如抗癌特性、预防幽门螺杆菌感染和促进免疫功能等（Yang et al.，2017）。最具代表性的 8 种 MFGM 蛋白是黏蛋白 1（MUC1）、黄嘌呤氧化还原酶（XO/XDH）、黏蛋白 15（MUC15 或 PASⅢ）、CD36（PASⅣ）、嗜乳脂蛋白（BTN）和乳凝集素（PASⅥ/Ⅶ）、脂肪分化相关蛋白（ADPH）和脂肪酸结合蛋白（FABP）（Riccio，2004）。MFGM 蛋白组成相当复杂，在 SDS-聚丙烯酰胺凝胶电泳（SDS-PAGE）凝胶中可观察到约 37 条蛋白带（Ye et al.，2002）。然而，MFGM 还有很多蛋白质，特别是低丰度蛋白，通常不能在 SDS-PAGE 电泳图像中观察到（Le et al.，2013）。近年来，研究人员利用液相色谱结合串联质谱（LC-MS/MS）技术的蛋白质组学法对 MFGM 蛋白的含量和种类进行了研究。用该方法从牛乳中鉴定了 MFGM 的 120 种蛋白质（Reinhardt and Lippolis，2006）。采用基于 LC-MS/MS 的检测方法，对 MFGM 蛋白含有的乳清蛋白质浓缩物中的 244 种蛋白质和酪蛋白浓缩物中的 133 种蛋白质进行了鉴定（Affolter et al.，2010）。利用 iTRAQ 偶联的 LC-MS/MS 可以对 411 种人和牛初乳 MFGM 蛋白进行定性和定量（Yang et al.，2017）。

3.6.4 维生素结合蛋白

牛乳中还含有许多其他微量的蛋白质，如维生素结合蛋白（Salter et al., 1972）。牛乳中的叶酸能够与一种微量乳清蛋白质紧密结合。约 50 年前，牛乳中的叶酸结合蛋白（FBP）被发现，这种结合蛋白的含量过高，具有与添加的叶酸结合的能力（Salter et al., 1972）。它通过抗氧化在叶酸的利用、储存和运输中发挥着重要作用（Svendsen et al., 1979），它能够促进哺乳动物对叶酸的吸收或防止分泌的叶酸被细菌利用（Birn et al., 2005）。蛋白质部分的分子质量为 25.7 kDa。结合叶酸的部分含有碳水化合物，最终分子质量约为 30 kDa，在 pH 7.4 时发生聚合（Pedersen et al., 1980）。该分子由单一的多肽链组成，含有 222 个氨基酸残基和 8 个二硫键（Svendsen et al., 1984）。牛乳中还有其他维生素结合蛋白，如维生素 B_{12} 结合蛋白和维生素 D 结合蛋白。

3.7 总　　结

乳清蛋白质是多种分泌蛋白的混合物。本章讨论了其主要蛋白质的化学特性、分离方法和功能特性，还讨论了生长因子等活性较高的微量蛋白质。毫无疑问，所有蛋白质的结构和功能决定了乳清蛋白质的性质和应用，这将在后面的章节中进行讨论。

参 考 文 献

Adams, J.J., Anderson, B.F., Norris, G.E. et al. (2006). Structure of bovine beta-lactoglobulin (variant A) at very low ionic strength. *Journal of Structural Biology*, **154** (3): 246-254.

Affolter, M., Grass, L., Vanrobaeys, F. et al. (2010). Qualitative and quantitative profiling of the bovine milk fat globule membrane proteome. *Journal of Proteomics* **73** (6): 1079-1088.

Aoki, K., Hiramatsu, K., Kimura, K. et al. (1969). Heat denaturation of bovine serum albumin. I. Analysis by acylamide-gel electrophoresis. *Bull. Inst. Chem. Res.* **47** (4): 274-282.

Arai, M. and Kuwajima, K. (1996). Rapid formation of a molten globule intermediate in refolding of α-lactalbumin. *Folding & Design*, **1** (4): 275-287.

Arunkumar, A. and Etzel, M.R. (2014). Fractionation of α-lactalbumin and β-lactoglobulin from bovine milk serum using staged, positively charged, tangential flow ultrafiltration membranes. *Journal of Membrane Science* **454** (6): 488-495.

Bhattacharjee, S., Bhattacharjee, C., and Datta, S. (2006). Studies on the fractionation of β-lactoglobulin from casein whey using ultrafiltration and ion-exchange membrane chromatography. *Journal of Membrane Science* **275** (1): 141-150.

Birn, H., Zhai, X., Holm, J. et al. (2005). Megalin binds and mediates cellular internalization of folate binding protein. *The FEBS Journal* **272** (17): 4423-4430.

Björck, L. (1978). Antibacterial effect of the lactoperoxidase system on psychotrophic bacteria in milk. *Journal of Dairy Research* **45** (1): 109-118.

Björck, L., Rosen, C., Marshall, V. et al. (1975). Antibacterial activity of the lactoperoxidase system in milk against pseudomonads and other gram-negative bacteria. *Applied Microbiology* **30** (2): 199-204.

Bogahawaththa, D., Chandrapala, J., and Vasiljevic, T. (2017). Thermal denaturation of bovine immunoglobulin G and its association with other whey proteins. *Food Hydrocolloids* **72**: 350-357.

Bolder, S.G., Vasbinder, A.J., Lmc, S. et al. (2007). Heat-induced whey protein isolate fibrils: conversion, hydrolysis, and disulphide bond formation. *International Dairy Journal* **17** (7): 846-853.

Boye, J.I. and Alli, I. (2000). Thermal denaturation of mixtures of α-lactalbumin and β-lactoglobulin: a differential scanning calorimetric study. *Food Research International* **33** (8): 673-682.

Bramaud, C., Aimar, P., and Daufin, G. (1997). Whey protein fractionation: isoelectric precipitation of alpha-lactalbumin under gentle heat treatment. *Biotechnology and Bioengineering* **56** (4): 391-397.

Brownlow, S., Morais Cabral, J.H., Cooper, R. et al. (1997). Bovine beta-lactoglobulin at 1.8 A resolution - still an enigmatic lipocalin. *Structure* **5** (4): 481-495.

Butler, J.E. (1969). Bovine immunoglobulins: a review. *Journal of Dairy Science* **52** (12): 1895-1909.

Carter, D.C. and Ho, J.X. (1994). Structure of serum albumin. *Advances in Protein Chemistry* **45**: 153-203. https://doi.org/10.1016/S0065-3233(08)60640-3.

Chaneton, L., Saez, J.P., and Bussmann, L.E. (2011). Antimicrobial activity of bovine β-lactoglobulin against mastitis-causing bacteria. *Journal of Dairy Science*, **94** (1): 138-145.

Chatterton, D.E.W., Smithers, G., Roupas, P. et al. (2006). Bioactivity of β-lactoglobulin and α-lactalbumin-technological implications for processing. *International Dairy Journal,* **16** (11): 1229-1240.

Chen, L., Guo, C., Guan, Y. et al. (2007). Isolation of lactoferrin from acid whey by magnetic affinity separation. *Separation and Purification Technology* **56** (2): 168-174.

Chen, C.C., Tu, Y.Y., and Chang, H.M. (2010). Thermal stability of bovine milk immunoglobulin G (IgG) and the effect of added thermal protectants on the stability. *Journal of Food Science* **65** (2): 188-193.

Chevalier, F., Chobert, J.M., Genot, C. et al. (2001). Scavenging of free radicals, antimicrobial, and cytotoxic activities of the Maillard reaction products of beta-lactoglobulin glycated with several sugars. *Journal of Agricultural and Food Chemistry* **49** (10): 5031-5038.

Considerations, I. (2006). Immunoglobulins-basic considerations. *Journal of Neurology* **253** (S5): 9-17.

Divsalar, A., Saboury, A.A., Ahmad, F. et al. (2009). Effects of temperature and chromium (III) ion on the structure of bovine β-lactoglobulin-A. *Journal of the Brazilian Chemical Society* **20** (10): 245-248.

Elias, R.J., And, M.C., and Decker, E.A. (2005). Antioxidant activity of cysteine, tryptophan, and methionine residues in continuous phase β-lactoglobulin in oil-in-water emulsions. *Journal of*

Agricultural and Food Chemistry **53** (26): 10248-10253.

Elias, R.J., Bridgewater, J.D., Vachet, R.W. et al. (2006). Antioxidant mechanisms of enzymatic hydrolysates of beta-lactoglobulin in food lipid dispersions. *Journal of Agricultural and Food Chemistry* **54** (25): 9565-9572.

Elloly, M.M. (2011). Composition, properties and nutritional aspects of milk fat globule membrane - a review. *Polish Journal of Food and Nutrition Sciences* **61** (1): 7-32.

El-Sayed, M.M. and Chase, H.A. (2011). Trends in whey protein fractionation. *Biotechnology Letters* **33** (8): 1501-1511.

Farrell, H.M. Jr., Jimenez-Flores, R., Bleck, G.T. et al. (2004). Nomenclature of the proteins of cows' milk--sixth revision. *Journal of Dairy Science* **87** (6): 1641-1674.

Fong, B.Y., Norris, C.S., and Macgibbon, A.K.H. (2007). Protein and lipid composition of bovine milkfat- globule membrane. *International Dairy Journal* **17** (4): 275-288.

Fox, K.K., Holsinger, V.H., Posati, L.P. et al. (1967). Separation of β-Lactoglobulin from other milk serum proteins by trichloroacetic acid. *Journal of Dairy Science* **50** (9): 1363-1367.

Fukumoto, L.R., Li-Chan, E., Kwan, L. et al. (1994). Isolation of immunoglobulins from cheese whey using ultrafiltration and immobilized metal affinity chromatography. *Food Research International* **27** (4): 335-348.

Gerberding, S.J. and Byers, C.H. (1998). Preparative ion-exchange chromatography of proteins from dairy whey. *Journal of Chromatography A* **808** (1-2): 141-151.

Guo, M.R. and Wang, G.R. (2016). Whey protein polymerisation and its applications in environmentally safe adhesives. *International Journal of Dairy Technology* **69** (4): 481-488.

Guo, M.R., Fox, P.F., Flynn, A. et al. (1995). Susceptibility of beta-lactoglobulin and sodium caseinate to proteolysis by pepsin and trypsin. *Journal of Dairy Science* **78** (11): 2336-2344.

Heine, W.E., Klein, P.D., and Reeds, P.J. (1991). The importance of β-lactalbumin in infant nutrition. *Journal of Nutrition* **121** (3): 277-283.

Hernández-Ledesma, B., Ramos, M., Recio, I. et al. (2006). Effect of β-Lg hydrolysis with thermolysin under denaturing temperatures on the release of bioactive peptides. *Journal of Chromatography A* **1116** (1): 31-37.

Hoffmann, M.A.M., Miltenburg, J.C.V., and Mil, P.J.J.M.V. (1997). The suitability of scanning calorimetry to investigate slow irreversible protein denaturation. *Thermochimica Acta* **306** (1-2): 45-49.

Hurley, W.L. and Theil, P.K. (2011). Perspectives on immunoglobulins in colostrum and milk. *Nutrients* **3** (4): 442-474.

Hwang, P.M., Zhou, N., Shan, X. et al. (1998). Three-dimensional solution structure of lactoferricin b, an antimicrobial peptide derived from bovine lactoferrin. *Biochemistry* **37** (12): 4288-4298.

Implvo, F., Pinho, O., Mota, M.V. et al. (2007). Preparation of ingredients containing an ace-inhibitory peptide by tryptic hydrolysis of whey protein concentrates. *International Dairy Journal* **17** (5): 481-487.

Jelen, P. (2010). Bioactive components in milk and dairy products. *International Dairy Journal* **20** (8): 560.

Jongh, H.H.J.D., Groneveld, T., and Groot, J.D. (2001). Mild isolation procedure discloses new protein structural properties of β-lactoglobulin. *Journal of Dairy Science* **84** (3): 562-571.

Kandagal, P.B., Ashoka, S., Seetharamappa, J. et al. (2006). Study of the interaction of an anticancer drug with human and bovine serum albumin: spectroscopic approach. *Journal of Pharmaceutical and Biomedical Analysis* **41** (2): 393-399.

Kinekawa, Y.I. and Kitabatake, N. (1996). Purification of β-lactoglobulin from whey protein concentrate by pepsin treatment. *Journal of Dairy Science* **79** (3): 350-356.

Klajnert, B. and Bryszewska, M. (2002). Fluorescence studies on PAMAM dendrimers interactions with bovine serum albumin. *Bioelectrochemistry* **55** (1): 33-35.

Konrad, G. and Kleinschmidt, T. (2008). A new method for isolation of native α-lactalbumin from sweet whey. *International Dairy Journal* **18** (1): 47-54.

Korhonen, H., Marnila, P., and Gill, H.S. (2000). Milk immunoglobulins and complement factors. *British Journal of Nutrition* **84** (S1): 75-80.

Kussendrager, K.D. and van Hooijdonk, A.C. (2000). Lactoperoxidase: physico-chemical properties, occurrence, mechanism of action and applications. *British Journal of Nutrition* **84** (S1): 19-25.

Kuwajima, K., Yamaya, H., and Sugai, S. (1996). The burst-phase intermediate in the refolding of betalactoglobulin studied by stopped-flow circular dichroism and absorption spectroscopy. *Journal of Molecular Biology* **264** (4): 806-822.

Le, T.T., Debyser, G., Gilbert, W. et al. (2013). Distribution and isolation of milk fat globule membrane proteins during dairy processing as revealed by proteomic analysis. *International Dairy Journal* **32** (2): 110-120.

Li, S.Q., Zhang, Q.H., Lee, Y.Z. et al. (2010). Effects of pulsed electric fields and thermal processing on the stability of bovine immunoglobulin G (IgG) in enriched soymilk. *Journal of Food Science* **68** (4): 1201-1207.

Liang, J., Yan, H., Yang, H.J. et al. (2016). Synthesis and controlled-release properties of chitosan/β-lactoglobulin nanoparticles as carriers for oral administration of epigallocatechin gallate. *Food Science and Biotechnology* **25** (6): 1583-1590.

Liu, J., Tian, J., He, W. et al. (2004). Spectrofluorimetric study of the binding of daphnetin to bovine serum albumin. *Journal of Pharmaceutical and Biomedical Analysis* **35** (3): 671-677.

Liu, X., Li, S., Jiang, X. et al. (2006). Conformational changes of β-lactoglobulin induced by anionic phospholipid. *Biophysical Chemistry* **121** (3): 218-223.

Liu, H.C., Chen, W.L., and Mao, S.J. (2007). Antioxidant nature of bovine milk beta-lactoglobulin. *Journal of Dairy Science* **90** (2): 547-555.

Lönnerdal, B. and Iyer, S. (1995). Lactoferrin: molecular structure and biological function. *Annual Review of Nutrition* **15** (1): 93-110.

Lozano, J.M., Giraldo, G.I., and Romero, C.M. (2008). An improved method for isolation of β-lactoglobulin. *International Dairy Journal* **18** (1): 55-63.

Lucena, M.E., Alvarez, S., Menendez, C. et al. (2006). Beta-lactoglobulin removal from whey protein concentrates: production of milk derivatives as a base for infant formulas. *Separation and Purification Technology* **52** (2): 310-316.

Madureira, A.R., Pereira, C.I., Gomes, A.M.P. et al. (2007). Bovine whey proteins - overview on their main biological properties. *Food Research International* **40** (10): 1197-1211.

Mao, X., Zhang, G.F., Li, C. et al. (2016). One-step method for the isolation of α-lactalbumin and β-lactoglobulin from cow's milk while preserving their antigenicity. *International Journal of Food Properties* **20** (4): 792-800.

Maruyama, S. and Suzuki, H. (1982). A peptide inhibitor of angiotensin I converting enzyme in the tryptic hydrolysate of casein. *Agricultural and Biological Chemistry* **46** (5): 1393-1394.

Mcintosh, G.H., Regester, G.O., Le Leu, R.K. et al. (1995). Dairy proteins protect against dimethylhydrazine-induced intestinal cancers in rats. *Journal of Nutrition* **125** (4): 809-816.

Mehra, R., Marnila, P., and Korhonen, H. (2006). Milk immunoglobulins for health promotion. *International Dairy Journal* **16** (11): 1262-1271.

Mezanieto, M.A., Gonzalezcordova, A.F., Becerrilperez, C.M. et al. (2012). Associations between variants A and B of β-lactoglobulin and milk production and composition of Holstein and milking tropical criollo cows. *Agrociencia* **46** (1): 15-22.

Michaela, M., Zatam, V., and Garya, R. (2008). Flavor of whey protein concentrates and isolates. *International Dairy Journal* **18** (6): 649-657.

Mietens, C., Keinhorst, H., Hilpert, H. et al. (1979). Treatment of infantile e. coli gastroenteritis with specific bovine anti-e. coli milk immunoglobulins. *European Journal of Pediatrics* **132** (4): 239-252.

Montilla, A., Casal, E., Moreno, F.J. et al. (2007). Isolation of bovine α-lactoglobulin from complexes with chitosan. *International Dairy Journal* **17** (5): 459-464.

Mullally, M.M., Meisel, H., and Fitzgerald, R.J. (1996). Synthetic peptides corresponding to β-lactalbumin and β-lactoglobulin sequence with angiotensin-I-converting enzyme inhibitory activity. *Biological Chemistry Hoppe-Seyler* **377** (4): 259-260.

Ng, S.K., Nyam, K.L., Nehdi, I.A. et al. (2016). Impact of stirring speed on β-lactoglobulin fibril formation. *Food Science and Biotechnology* **25** (1): 15-21.

Oevermann, A., Engels, M., Thomas, U. et al. (2003). The antiviral activity of naturally occurring proteins and their peptide fragments after chemical modification. *Antiviral Research* **59** (1): 23-33.

Pakkamen, R. and Aalto, J. (1997). Growth factors and antimicrobial factors of bovine colostrum. *International Dairy Journal* **7** (5): 285-297.

Pan, Y., Shiell, B., Wan, J. et al. (2007). The molecular characterisation and antimicrobial properties of amidated bovine-lactoglobulin. *International Dairy Journal* **17** (12): 1450-1459.

Papadopoulou, A., Green, R. J., and Frazier, R. A. (2005). Interaction of flavonoids with bovine serum albumin: a fluorescence quenching study. *Journal of Agricultural and Food Chemistry* **53**(1): 158-163.

Paul, B.K., Ghosh, N., and Mukherjee, S. (2014). Binding interaction of a prospective chemotherapeutic antibacterial drug with β-lactoglobulin: results and challenges. *Langmuir the ACS Journal of Surfaces and Colloids* **30** (20): 5921-5929.

Pedersen, T.G., Svendsen, I., Hansen, S.I. et al. (1980). Aggregation of a folate-binding protein from cow's milk. *Carlsberg Research Communications* **45** (2): 161.

Pellegrini, A. (2003). Antimicrobial peptides from food proteins. *Current Pharmaceutical Design* **9** (16): 1225.

Pellegrini, A., Bramaz, N.P., Von, F.R. et al. (1999). Isolation and identification of three bactericidal domains in the bovine alpha-lactalbumin molecule. *Biochimica et Biophysica Acta (BBA)-General Subjects* **1426** (3): 439-448.

Pellegrini, A., Dettling, C., Thomas, U. et al. (2001). Isolation and characterization of four bactericidal domains in the bovine beta-lactoglobulin. *Biochimica et Biophysica Acta (BBA) General Subjects* **1526** (2): 131-140.

Permyakov, E.A. and Berliner, L.J. (2000). α-Lactalbumin: structure and function. *FEBS Letters* **473** (3): 269-274.

Phillips, G.O. and Williams, P.A. (2011). Whey proteins. In: *Handbook of food proteins. Handbook of food proteins*. Woodhead Publishing: Cambridge, 30-55.

Pierce, A., Colavizza, D., Benaissa, M. et al. (1991). Molecular cloning and sequence analysis of bovine lactotransferrin. *European Journal of Biochemistry* **196** (1): 177-184.

Pihlanto-Leppälä, A. (2000). Bioactive peptides derived from bovine whey proteins: opioid and aceinhibitory peptides. *Trends in Food Science and Technology* **11** (9-10): 347-356.

Pihlantoleppala, A., Rokka, T., and Korhonen, H. (1998). Angiotensin I converting enzyme ryaninhibitory peptides derived from bovine milk proteins. *International Dairy Journal* **8** (8): 325-331.

Rabiey, L. and Britten, M. (2009). Effect of protein composition on the rheological properties of acidinduced whey protein gels. *Food Hydrocolloids* **23** (3): 973-979.

Reddy, I.M., Kella, N.K.D., and Kinsella, J.E. (1988). Structural and conformational basis of the resistance of β-lactoglobulin to peptic and chymotryptic digestion. *Journal of Agricultural & Food Chemistry* **36** (4): 737-741.

Redfield, C., Schulman, B.A., Milhollen, M.A. et al. (1999). Alpha-lactalbumin forms a compact molten globule in the absence of disulfide bonds. *Nature Structural Biology* **6** (10): 948-952.

Reinhardt, T.A. and Lippolis, J.D. (2006). Bovine milk fat globule membrane proteome. *Journal of Dairy Research* **73** (4): 406-416.

Riccio, P. (2004). The proteins of the milk fat globule membrane in the balance. *Trends in Food Science and Technology* **15** (9): 458-461.

Salter, D., Ford, J., Scott, K. et al. (1972). Isolation of the folate-binding protein from cow's milk by the use of affinity chromatography. *Febs Letters* **20** (3): 302-306.

Santos, M.J., Teixeira, J.A., and Rodrigues, L.R. (2012). Fractionation of the major whey proteins and isolation of β-lactoglobulin variants by anion exchange chromatography. *Separation and Purification Technology* **90**: 133-139.

Schlatterer, B., Baeker, R., and Schlatterer, K. (2004). Improved purification of beta-lactoglobulin from acid whey by means of ceramic hydroxyapatite chromatography with sodium fluoride as a displacer. *Journal of Chromatography B* **807** (2): 223-228.

Schokker, E.P., Singh, H., Pinder, D.N. et al. (1999). Characterization of intermediates formed during heat-induced aggregation of beta-lactoglobulin AB at neutral pH. *International Dairy Journal* **9** (11): 791-800.

Schokker, E.P., Singh, H., and Creamer, L.K. (2000). Heat-induced aggregation of β-lactoglobulin A and B with α-lactalbumin. *International Dairy Journal* **10** (12): 843-853.

Seifu, E., Buys, E.M., and Donkin, E.F. (2005). Significance of the lactoperoxidase system in the dairy industry and its potential applications: a review. *Trends in Food Science and Technology* **16** (4): 137-154.

Sharp, J.S., Becker, J.M., and Hettich, R.L. (2004). Analysis of protein solvent accessible surfaces by photochemical oxidation and mass spectrometry. *Analytical Chemistry* **76** (3): 672-683.

Sklar, L.A., Hudson, B.S., and Simoni, R.D. (1977). Conjugated polyene fatty acids as fluorescent probes: binding to bovine serum albumin. *Biochemistry* **16** (23): 5100-5108.

Smithers, G.W., Ballard, F.J., Copeland, A.D. et al. (1996). Symposium: advances in dairy foods processing and engineering : new opportunities from the isolation and utlization of whey proteins. *Journal of Dairy Science* **79**: 1454-1459.

Soave, R. and Armstrong, D. (1987). Cryptosporidium and cryptosporidiosis. *Reviews of Infectious Diseases* **8**: 1012-1023.

Sørensen, M. and Sørensen, S.P.L. (1939). The proteins in whey. *C. R. Trav. Lab. Carlsberg* **23**: 55-99.

Stănciuc, N. and Râpeanu, G. (2010). An overview of bovine α-lactalbumin structure and functionality. *Annals of the University Dunarea De Jos of Galati:fascicle VI Food Technology* **34** (2): 82-93.

Svendsen, I., Martin, B., Pedersen, T.G. et al. (1979). Isolation and characterization of the folatebinding protein from cow's milk. *Carlsberg Research Communications* **44** (2): 89.

Svendsen, I., Hansen, S.I., Holm, J. et al. (1984). The complete amino acid sequence of the folatebinding protein from cow's milk. *Carlsberg Research Communications* **49** (1): 123-131.

Tacket, C.O., Losonsky, G., and Link, H. (1988). Protection by milk immunoglobulin concentrate against oral challenge with enterotoxigenic Escherichia coil. *New England Journal of Medicine* **318**: 1240-1241.

Tacket, C.O., Binion, S.B., Bostwick, E. et al. (1992). Efficacy of bovine milk immunoglobulin concentrate in preventing illness after Shigella flexneri challenge. *American Journal of Tropical Medicine and Hygiene* **47**: 276-283.

Tzipori, S., Roberton, D., and Chapman, C. (1986). Remission of diarrhea due to cryptosporidiosis in an immunodeficient child treated with hyperimmune bovine colostrum. *British Medical Journal* **293** (9): 1276-1277.

Walsh, H. (2014). Functional properties of whey protein and its application in nanocomposite materials and functional foods. Doctoral dissertation. The University of Vermont.

Wassef, B.O., Sylvief, G., Sylviel, T. et al. (2008). Separation of minor protein components from whey protein isolates by heparin affinity chromatography. *International Dairy Journal* **18** (10): 1043-1050.

Wolfson, L.M. and Sumner, S.S. (1993). Antibacterial activity of the lactoperoxidase system: a review. *Journal of Food Protection* **56** (10): 887-892.

Xu, Y., Sleigh, R., Hourigan, J. et al. (2000). Separation of bovine immunoglobulin G and glycomacropeptide from dairy whey. *Process Biochemistry* **36** (5): 393-399.

Yadav, J.S., Yan, S., Pilli, S. et al. (2015). Cheese whey: a potential resource to transform into bioprotein, functional/nutritional proteins and bioactive peptides. *Biotechnology Advances* **33** (6): 756-774.

Yang, M.C., Guan, H.H., Liu, M.Y. et al. (2010). Crystal structure of a secondary vitamin d3 binding site of milk beta-lactoglobulin. *Proteins Structure Function and Bioinformatics* **71** (3): 1197-1210.

Yang, M., Peng, X., Wu, J. et al. (2017). Differential proteomic analysis of milk fat globule membrane proteins in human and bovine colostrum by iTRAQ-coupled LC-MS/MS. *European Food Research and Technology* **243** (5): 901-912.

Ye, X.Y., Yoshida, S., and Ng, T.B. (2000). Isolation of lactoperoxidase, lactoferrin, α-lactalbumin, β-lactoglobulin B and β-lactoglobulin A from bovine rennet whey using ion exchange chromatography. *The International Journal of Biochemistry & Cell Biology* **32** (11-12): 1143-1150.

Ye, A., Singh, H., Taylor, M.W. et al. (2002). Characterization of protein components of natural and heat-treated milk fat globule membranes. *International Dairy Journal* **12** (4): 393-402.

Yong, S.C., Song, K.B., and Yamda, K. (2010). Effect of ultraviolet irradiation on molecular properties and immunoglobulin production-regulating activity of beta-lactoglobulin. *Food Science and Biotechnology* **19** (3): 595-602.

第4章 乳清蛋白质的结构和变性以及与其他食品组分的相互作用

Cuina Wang[1] and Mingruo Guo[2,3]

1. Department of Food Science, College of Food Science and Engineering, Jilin University, ChangChun, People's Republic of China
2. Department of Nutrition and Food Science, University of Vermont, Burlington, USA
3. College of Food Science, Northeast Agriculture University, Harbin, People's Republic of China

4.1 乳清蛋白质的结构和变性

乳清蛋白质（WP）是一类具有高级结构的球状蛋白质，包含一级、二级、三级甚至四级结构。在标准pH、离子强度和温度等条件下，乳清蛋白质以一种特殊的构象存在，这种构象称为"天然状态"。化学、酶或物理（加热、辐射以及超声）处理可引起构象和理化性质的改变，这种变化称为蛋白质变性。乳清蛋白质的变性包含两个步骤：首先，天然蛋白质经处理后呈活化状态或分子结构展开；随后，展开的分子发生（不可逆性）聚集（Singh and Havea, 2003）。蛋白质特殊的功能特性是由其特定的构象状态所决定的，任何状态上的改变都会影响该蛋白质的功能。虽然乳清蛋白质变性会引起溶解度下降从而导致功能特性改变，但也能产生一些有益的功能特性，如改善表面性质（Moro et al., 2001）。

4.1.1 热变性

热处理是乳品工业中常用的加工技术。加热能引起溶液中乳清蛋白质分子运动加剧，破坏了一些稳定天然构象的作用力，如氢键、疏水作用（Boye et al., 1997a, b），蛋白质分子展开并暴露出活性氨基酸（非极性残基及半胱氨酸）（Gracia-Julia et al., 2008），这些活性氨基酸在一定pH条件下引起乳清蛋白质分子聚集。

1. 加热引起β-乳球蛋白结构变化

β-乳球蛋白（β-LG）的含量占总乳清蛋白质一半左右。室温、pH 5.5～7

条件下，β-LG 主要以二聚体的形式存在（与其在牛乳中的浓度相同）（Verheul et al., 1998）。当温度超过 40℃或在极端 pH 条件下，β-LG 二聚体将解离为单体（Leeb et al., 2017）。β-LG 单体含有两个二硫键（Cys66-Cys160 和 Cys106-Cys119）以及一个游离巯基（Cys121）。温度超过 60℃时，β-LG 的三级结构展开，疏水基团和巯基暴露。温度在 65～70℃时，β-LG 发生可逆性的构象变化（De Wit and Swinkels, 1980）。温度达到 70℃以上时，蛋白质变性聚集形成低聚体，低聚体在临界浓度下或盐的作用下进一步形成较大的聚集体。内部 Cys121 的游离巯基暴露引发巯基-二硫键交换反应，导致不可逆的聚集/聚合（Galani and Owusu Apenten, 2010），因此热变性过程是不可逆的（Gauche et al., 2010）。在复原脱脂乳中，β-LG 表现出不同的变性性质，在较宽的温度范围（75～100℃）内，β-LG（总固形物浓度 9.6%～38.4%）的热变性反应级数为 1.5（Anema, 2000）。

β-LG A 与β-LG B 的热稳定性不同，在 pH 3～9、浓度为 10%时，β-LG B 比β-LG A 热稳定性更好（Boye et al., 1997a, b）。β-LG A 和β-LG B 在热聚集反应中也表现出不同的行为，如聚集速率以及形成的聚集体的大小。

2. 乳清蛋白质的热变性

乳清蛋白质是一类球蛋白的混合物，β-LG 对乳清蛋白质变性影响很大。β-LG 的变性温度（T_d）是 71.9℃，而当β-LG 与α-LA 以 1:1 混合时，变性温度降低至 69.1℃（Boye and Alli, 2000）。α-LA 的存在降低了β-LG 亚型 A 和 B 中较小聚集体的比例，同时增加了较大聚集体的数量（Schokker et al., 2000）。80℃下，4%的β-LG 在 100 mmol/L 磷酸钾缓冲液（pH 6.8）中加热 30 min，能形成可倒置的凝胶，但向 2%的β-LG 中加入 6%的α-LA 就可以显著提高凝胶硬度。通过巯基-二硫键交换反应，β-LG 和α-LA 相互作用形成的可溶性聚集物有利于凝胶的形成和稳定（Matsudomi et al., 1992），BSA 也能与β-LG 之间形成二硫键，从而提高凝胶的硬度（Havea et al., 2001）。

β-LG 能提高α-LA 的热稳定性（变性温度可提高 2.5℃）（Boye and Alli, 2000），由于α-LA 没有游离巯基，不能通过加热形成聚集体，即使 90℃下α-LA（1.5%）加热 24 min，也未能检测到聚集体的产生。但在α-LA 溶液中加入β-LG 或 BSA 则会使α-LA 发生聚集反应，而且这种聚集的速度和程度取决于其他蛋白质中游离巯基的浓度（Calvo et al., 1993）。在α-LA 存在的条件下，有活性巯基的β-LG 能从α-LA 中夺走一个巯基，形成二硫键。因此，结构伸展的α-LA 就可以发生相互作用，最终形成β-LG 聚集体、α-LA 聚集体和β-LG/α-LA 聚集体的混合物（Vardhanabhuti and Foegeding, 1999）。

4.1.2 乳清蛋白质的酶促变性

1. 乳清蛋白质的水解

β-LG 是乳清中最主要的蛋白质，但它也是婴儿食品中的主要变应原（配料中的乳清蛋白质产品含有β-LG）。β-LG 结构紧凑，难被酶解，在胃部的消化速度比其他乳蛋白缓慢得多。婴儿肠道比较脆弱，一些完整的蛋白质分子被肠道吸收后会引起过敏反应，减少牛乳致敏性最常见的方法是用各种蛋白酶水解乳蛋白（Lee et al.，2001）。

2. 乳清蛋白质的交联

酶催化的蛋白交联常用于乳清蛋白质的改性。转谷氨酰胺酶（TG）、脂肪氧化酶、赖氨酸氧化酶、过氧化物酶和漆酶都可以催化乳清蛋白质分子之间的交联反应。TG 是应用最广泛的一种，它是一种存在于真核细胞和微生物中的酶（Schmid et al.，2015）。TG 催化酰基转移反应，连接谷氨酰胺和赖氨酸残基侧链，生成ε-(γ-谷氨酰)赖氨酸键，释放氨（Eissa et al.，2004）。

β-LG 分子含有 4 个谷氨酰胺和 15 个赖氨酸残基，而α-LA 分子含有 5 个谷氨酰胺和 12 个赖氨酸残基。与 TG 发生交联反应的前提是充分暴露底物蛋白的谷氨酰胺和赖氨酸残基。天然β-LG 的结构紧密，能抵制 TG 的作用，因此酶交联反应需要结构伸展的β-LG（Stender et al.，2018）。二硫键是维持天然β-LG 三级结构的主要作用力，二硫苏糖醇（DTT）等变性剂（Eissa et al.，2004）能破坏二硫键导致β-LG 变性，分子结构伸展，有利于 TG 的催化交联。DTT 不允许在食品中应用，可以使用其他还原剂如半胱氨酸和亚硫酸盐，但其反应速率远低于 DTT（DeJong and Koppelman，2002）。

天然α-LA 对 TG 交联不敏感。钙离子能使α-LA 结构处于稳定状态，去除钙离子后α-LA 的天然结构被破坏，导致其转变为熔融的球体状态。EDTA 能够螯合α-LA 中的钙离子使其容易被细菌 TG 交联（DeJong and Koppelman，2002）。

4.1.3 乳清蛋白质的超声改性

超声波是由频率超过人类听觉极限（约 20 kHz）的声波组成。通过调节频率，超声波可以在食品加工等许多工业上应用。超声波可分为低强度和高强度两种，低强度超声的频率大于 100 kHz，强度小于 1 W/cm^2，无创、无损，高强度超声（强度大于 1 W/cm^2，频率在 20～100 kHz 之间）已被广泛使用，具有用时短、操作简单、易于控制、能耗低等优点（Ma et al.，2018）。高强度超声可以对食品的物理、化学或生化特性产生影响。超声波产生热量和空化效应，热量是由探针、介质和反应器壁之间的摩擦产生的，空化效应是指气泡的形成和破裂，同时在空化气泡中心产生极高的压力和温度，这是超声化学中产生化学活性的主要机制

(Gordon and Pilosof, 2010)。

超声处理（24 kHz, 300 W/cm²）显著降低了 WPI 变性焓, 但其二级结构没有发生变化（Frydenberg et al., 2016）。超声处理（20 kHz, 振幅为 50%）WPC（5%）5 min 后, 其变性焓降低, 将处理时间延长到 60 min 时, 由于蛋白质聚集, 其变性焓有所升高。超声处理没有改变蛋白质的疏基含量, 但对蛋白质的二级结构和疏水性有微小的改变（Chandrapala et al., 2011）。20 kHz 下超声处理β-LG 60 min 后, β-LG 的活性疏基含量和表面疏水性持续增加, 表明二聚体结构裂解, 二级和三级结构发生微小变化（Chandrapala et al., 2012）。高强度超声（20 kHz, 60 W/cm²）处理β-LG, 发现β-LG 对胃蛋白酶和胰蛋白酶水解更敏感（Ma et al., 2018）。超声处理对α-LA 的影响更明显, 其表面疏水性得到显著增强（Chandrapala et al., 2012）。超声处理对 BSA 结构变化的影响很小, 但在碱性（pH>9）条件下, 其表面电荷明显增加、粒径增大、游离疏基数量减少, 这些变化可能与蛋白质聚集体的形成有关。圆二色谱和傅里叶变换红外光谱（FTIR）分析表明, BSA 的表面疏水性增强, 蛋白质的二级结构发生了变化（Gülseren et al., 2007）。

强度高、频率低的超声波会产生强大的剪切力和机械力, 这一特性已经被应用于降低聚合物和大分子的分子量。高强度超声（20 kHz）处理 85℃下预热 30 min 的 WPI（10%）5～40 min, 结果显示蛋白质的粒径均有所减小（Shen et al., 2017a, b）。经超声处理后, 乳清蛋白质-桃柁酚纳米颗粒粒径由 31 nm 减小到 24 nm（Ma et al., 2017）。超声（频率为 20 kHz）处理 WPI（10%）15 min 和 30 min 后, 所有样品中均出现蛋白粒径减小、分布变窄、自由比表面积明显增加的现象（Jambrak et al., 2014）。低频超声产生强大的剪切力, 高频超声产生大量的自由基, 空化过程中产生的物理剪切力是造成这一现象的主要原因（Ashokkumar, 2015）。

4.1.4 乳清蛋白质的辐射改性

电离能如伽马射线（钴-60 或铯-137）已应用于食品加工。一定剂量的伽马射线辐照可引起食物成分特性发生变化（Wang et al., 2018）。伽马射线辐照固体食物时, 分子直接吸收辐射能量并发生变化。辐照水溶液时, 水会产生羟基自由基和水合电子, 它们会与其他分子反应形成共价键, 这些化学变化大部分与热处理产生的变化相似（Chawla et al., 2009）。伽马射线辐照可以改善乳清蛋白质的理化性质, 辐照过程中形成的自由基（羟基自由基）促进两个相邻的酪氨酸分子结合形成双酪氨酸（Sabato et al., 2007）。与酪蛋白相比, 在 32 kGy 下辐照 WPC 时, 其分子量的变化很小, 原因是 WPC 的酪氨酸含量低于酪蛋白（Vachon et al., 2000）。由于辐照能产生交联效应, 它可以部分或全部替代加热处理（Cieśla et al., 2006）。

水分活度分别为 0.22、0.53、0.74 的β-LG 粉在剂量为 1～50 kGy 的辐照下, 并未产生结构上的变化, 原因是其拥有致密的结构。辐照处理β-LG 溶液（3 mg/mL

和 10 mg/mL）时，其三级和四级结构则出现了变化（Oliveira et al., 2007）。辐照处理高浓度的乳清蛋白质溶液（10%和 30%）时，其分子间通过形成二硫键发生交联（Wang et al., 2018）。用傅里叶变换红外光谱分析了辐照交联蛋白的特性，与未经辐照组相比，辐照蛋白的谱带强度发生变化，且变化与β折叠结构的增加、α螺旋和无规则结构的减少有相关性（Vu et al., 2012）。X 射线衍射分析表明，辐照改变了蛋白质的构象，使其变得更有序、更稳定（Le Tien et al., 2000）。用钴-60 射线以 0 kGy 和 32 kGy 剂量辐照溶液，由于交联作用，溶液黏度增加。将经过 32 kGy 辐照后的溶液进行加热，发现其黏度低于未经辐照的溶液（Cieśla et al., 2004）。辐照处理改变了能与 IgE 结合的抗原表位区域的结构，从而降低了不同食品过敏原的变应性（Tammineedi et al., 2013），3~10 kGy 剂量的辐照可降低牛奶过敏原的变应性（Lee et al., 2001）。

4.2 巯基和二硫键在乳清蛋白质聚集和凝胶中的作用

半胱氨酸有一个巯基（—SH），它可以通过氧化反应与其他巯基形成二硫键（—S—S—）。游离的巯基也可以与二硫键进行巯基-二硫键交换（Bryant and McClements, 1998; Visschers and de Jongh, 2005）。

巯基和二硫键在热凝胶和冷凝胶（包括盐诱导凝胶和酸诱导凝胶）中的作用如图 4.1 所示。热凝胶是通过加热浓度足够高的乳清蛋白质溶液形成的。冷凝胶的形成需要两个步骤，首先在无盐条件下加热天然蛋白质溶液制备蛋白质聚集体，然后在溶液冷却至室温后，降低 pH 产生凝胶，巯基在这些反应中起着重要的作用。

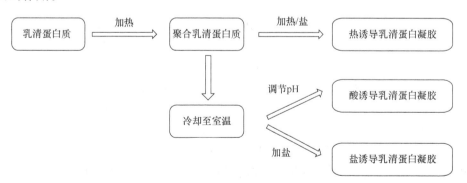

图 4.1 乳清蛋白质热/冷凝胶的形成过程

4.2.1 巯基和二硫键在乳清蛋白质聚集中的作用

巯基-二硫键交换反应形成的分子间二硫键在乳清蛋白质热聚集中起主要作用。

天然乳清蛋白质中的游离巯基和二硫键位于折叠分子的内部（Bryant and McClements，1998）。乳清蛋白质分子伸展暴露出的活性巯基[特别是β-LG 的半胱氨酸残基（氨基酸残基 121）上的巯基]对蛋白聚集起着重要作用。

当巯基阻断剂 N-乙基顺丁烯二酰亚胺（NEM）存在时，它可与巯基反应抑制蛋白质分子之间巯基-二硫键的交换反应，当 NEM 和 β-LG 的物质的量比为 1∶1 时，所有巯基都被阻断。浓度为 1%~5% 的 β-LG 在中性 pH、温度 65℃下加热时，也不会形成含二硫键的聚集体（Hoffmann and van Mil，1997）。α-LA 中没有游离巯基，所以当没有 β-LG 或 BSA 等含有巯基的蛋白质与之共存的时候，其自身在加热时无法形成聚集体（Roefs and Kruif，1994）。乳清蛋白质通过形成分子间二硫键实现聚合（Monahan et al.，1995）。完全暴露于水中的巯基（pK_a 在 8.3 左右）反应性在温和的酸性条件下大大降低（Visschers and de Jongh，2005）。pH 为 9 和 11、室温（22℃）下，聚合反应即可发生（SDS-PAGE 检测），而在 pH 3、5 和 7 时，需要分别加热至 85℃、75℃和 70℃时才能发生明显的聚合反应（Monahan et al.，1995）。

4.2.2 巯基和二硫键在乳清蛋白质凝胶中的作用

凝胶是一种类似固体的黏弹性物质，其三维网络结构中充满溶质。在高压、加热、剪切等条件下，乳清蛋白质能够形成凝胶。

巯基和二硫键在乳清蛋白质三维网络的形成中起重要作用。通过 DTT 的还原作用可以形成新的巯基并且可能形成新的分子间二硫键提高凝胶强度，但是 DTT 也可能减少网络结构中的二硫键而降低凝胶强度（Zirbel and Kinsella，1988）。β-LG 单体和 BSA 单体各有一个游离巯基，它们可以形成热凝胶。BSA（4%）和 β-LG（5%）在 90℃下加热 15 min 可形成凝胶。浓度分别为 2 mmol/L 和 5 mmol/L 的 DTT 提高了 β-LG 和 BSA 凝胶硬度，更高浓度的 DTT 会显著降低 BSA 和 β-LG 的凝胶硬度（Matsudomi et al.，1991）。在热诱导的 WPC 凝胶（80℃加热 30 min 或 85℃加热 60 min）中，分子间的二硫键似乎赋予了乳清蛋白质凝胶坚韧的橡胶特性，而非共价键则赋予了它们硬脆的质地（Havea et al.，2004）。

乳清蛋白质凝胶的初始微观结构主要是由酸诱导的非共价键决定的，在凝胶过程中形成的二硫键发挥了稳定网络结构和增加凝胶强度的作用（Alting et al.，2000）。酸诱导乳清蛋白质凝胶的硬度是由巯基的数目决定的，而不是由聚集体大小或其他结构特征决定的（Alting et al.，2003）。其他人的研究也证实了酸诱导凝胶的强度降低与巯基的含量有关。在次硫氰酸根离子和过氧化氢存在时，葡萄糖酸内酯（GDL）诱导的 β-LG 凝胶强度会降低，这是因为分子间二硫键的形成被抑制了（Hirano et al.，1999）。超声处理 GDL 诱导乳清蛋白质形成凝胶网络过程中，氧化巯基形成二硫键（2SH$\longrightarrow$S—S）起着至关重要的作用（Shen et al.，2017a，b）。

4.3 乳清蛋白质和酪蛋白的相互作用

酪蛋白是牛乳中的主要蛋白质成分,约占牛乳总蛋白质的80%。酪蛋白以超分子聚集体的形式存在于天然乳中,称为酪蛋白胶束或微粒。酪蛋白分子分为四种,分别是α_{s1}-酪蛋白、α_{s2}-酪蛋白、β-酪蛋白和κ-酪蛋白,它们相对含量大约是4:1:3.5:1.5(Dalgleish and Corredig,2012)。它们通过疏水作用、氢键和磷酸钙纳米簇盐桥结合在一起,并通过κ-酪蛋白的聚合电解质层来保持稳定(Yazdi and Corredig,2012)。微粒的许多特性都取决于胶束表面的κ-酪蛋白(Dalgleish,2011)。

4.3.1 模型系统中乳清蛋白质和酪蛋白的相互作用

α_s-酪蛋白表现出类似分子伴侣的性质,倾向于自结合成胶束状聚集体。在5%的乳清蛋白质中加入5%的α-酪蛋白可以将乳清蛋白质的变性温度降低2~3℃(Paulsson and Dejmek,1990)。α_s-酪蛋白不仅能防止巨大的不溶性聚集体形成,而且能抑制尺寸较大的可溶性聚集体积聚(Bhattacharyya and Das,1999)。将β-LG(6%)和α_s-酪蛋白(2%)混合物的pH调至6,70℃下加热20 min即可观察到上述抑制作用(Yong and Foegeding,2008)。

β-酪蛋白的分子伴侣性质使其成为目前研究最广泛的酪蛋白之一。β-酪蛋白可能通过疏水作用被固定在胶束中,因为β-酪蛋白有典型的疏水性。在乳清蛋白质和酪蛋白的混合物中,β-酪蛋白没有对变性温度产生影响(Paulsson and Dejmek,1990)。但是从β-酪蛋白降低了β-LG、α-LA和BSA溶液的浊度分析,β-酪蛋白改变了热诱导聚集作用(Kehoe and Foegeding,2010)。在pH为6.0、70~90℃加热20 min的条件下,添加大于等于0.05%的β-酪蛋白会导致β-LG溶液(6%)浊度下降。在90℃下,将pH为6.0的β-LG与β-酪蛋白(2%)混合物加热90 min可得到透明的溶液(Yong and Foegeding,2008)。这种变化取决于pH和离子强度,pH为6时浊度的降低程度最大,添加$CaCl_2$比添加NaCl的影响更大(Kehoe and Foegeding,2010)。

κ-酪蛋白对α-LA和BSA的变性没有影响,但它能使β-LG的变性温度降低3℃(Paulsson and Dejmek,1990)。κ-酪蛋白也表现出类似分子伴侣的活性,这一点可以通过乳清蛋白质溶液加热后浊度的降低来证明(Guyomarc'h et al.,2009)。

4.3.2 乳清蛋白质和酪蛋白微粒的相互作用

κ-酪蛋白含有半胱氨酸,加热过程中,β-乳球蛋白的游离巯基活性增加,能与其他变性乳清蛋白质以及酪蛋白微粒表面的κ-酪蛋白发生巯基-二硫键交换反应,这个过程导致酪蛋白微粒表面结构变化,改变了酪蛋白微粒性质,如增强了结合姜黄素的能力(Yazdi and Corredig, 2012)。

当加热至温度高于主要乳清蛋白质变性温度时,乳中κ-酪蛋白和乳清蛋白质之间就会形成颗粒(Donato et al., 2007)。因此,加热后的乳由被乳清蛋白质包裹的酪蛋白微粒和可溶性的乳清蛋白质聚集体组成,一部分热诱导乳清蛋白质/酪蛋白复合物吸附在酪蛋白微粒表面,另一部分则以可溶性复合物的形式存在于乳清相中(Guyomarc'h et al., 2010; Pesic et al., 2014; Qi et al., 1995; Vasbinder et al., 2003)。酪蛋白微粒与乳清蛋白质的相互作用在某些乳制品中起着重要作用。加热乳与未加热乳相比,其酸凝胶性质存在明显差异,如成胶 pH 改变、凝胶强度增加以及凝胶微观结构不同(Vasbinder et al., 2004)。未加热乳的酪蛋白微粒在 pH 4.9 时絮凝。乳清蛋白质等电点是 5.2,表面被包裹的酪蛋白微粒在较高的 pH 下才会絮凝(Vasbinder et al., 2004),随着复合物等电点的升高,乳的成胶 pH 也显著升高(Morand et al., 2012)。加热乳的成胶 pH 变化与变性乳清蛋白质在乳清蛋白质聚集体和乳清蛋白质-酪蛋白微粒复合物中的分布呈线性关系(Vasbinder et al., 2003)。显然,κ-酪蛋白的存在延缓了乳清蛋白质聚合物的形成,并促进了乳中稳定颗粒的形成(Donato et al., 2007)。在利用酶凝酪蛋白产生凝胶过程中,提前添加乳清蛋白质的比没有添加乳清蛋白质的样品所形成的凝胶具有更高的储能模量和形变模量(Solar and Gunasekaran, 2010)。乳清蛋白质复合物与酪蛋白微粒表面的κ-酪蛋白相互作用有利于酸凝胶形成,因此热聚合乳清蛋白质的加入提高了凝胶强度(Morand et al., 2011)。

关于乳清蛋白质和酪蛋白复合物的形成,有三种可能的途径(Donato and Guyomarc'h, 2009):①乳清蛋白质分子先在乳清中聚集,然后吸附在酪蛋白微粒上;②乳清蛋白质分子没有形成聚集体,而是单独地吸附在酪蛋白微粒上;③κ-酪蛋白分子游离在乳清中,与乳清蛋白质形成κ-酪蛋白/乳清蛋白质复合物,然后吸附到酪蛋白微粒上。

4.4 乳清蛋白质和糖类的相互作用

4.4.1 乳清蛋白质和糖类之间的美拉德反应

美拉德反应是碳水化合物还原端与蛋白质伯胺之间的缩合反应,常发生在食品的热加工过程中(Sun et al., 2012)。赖氨酸和精氨酸的ε-氨基或 N 端氨基酸使乳清蛋白质成为美拉德反应底物,β-乳球蛋白含 16 个活性氨基基团,其中包括 N 端和 15 个赖氨酸残基(Morgan et al., 1999),α-LA 的末端有 1 个氨基和 12 个赖氨酸残基(Wooster and Augustin, 2007)。

1. 乳清蛋白质和糖类的干热法美拉德反应

美拉德反应可分为干热和湿热两种类型。干热反应水分少，反应可控（Nasirpour et al.，2006），但其反应时间长、条件苛刻、可能形成不溶性物质。这种方法产生轻微的色泽变化使之适合食品加工。蛋白质与糖类的美拉德反应在空间上阻碍了蛋白质聚集（Akhtar and Dickinson，2007；Liu and Zhong，2013）。在 pH 3.0~7.0、NaCl 或 $CaCl_2$ 浓度为 0~150 mmol/L、88℃温度下加热 2 min，乳清蛋白质与麦芽糊精反应在加热前后都可以防止蛋白质的聚集。糖基化降低了乳清分离蛋白的等电点，提高了其变性温度（Liu and Zhong，2013），也可显著提高蛋白质的溶解度、热稳定性和乳化性能（Dai et al.，2015）。

1）葡聚糖

葡聚糖是由葡萄糖残基通过α（1→6）糖苷键连接而成的线形多糖（Sun et al.，2012），呈电中性，能阻止静电络合物形成，广泛应用于蛋白质偶联。葡聚糖对乳清蛋白质的糖基化作用提高了乳清蛋白质在酸性 pH 条件下的溶解度，如提高了β-LG 和 BSA 的热稳定性（Jiménez-Castaño et al.，2007）。葡聚糖与乳清蛋白质的糖基化程度取决于蛋白质/葡聚糖的类型、比例、反应条件等。为避免美拉德反应进入高级阶段，高分子质量葡聚糖（43 kDa）与β-LG 的最优糖基化条件为 60℃、水活度 0.44、葡聚糖与β-LG 的质量比 2∶1，在此条件下可生成最多的糠氨酸化合物（Jiménez-Castaño et al.，2005）。葡聚糖（10 kDa 和 20 kDa）与三种乳清蛋白质糖基化程度高低依次为 BSA＞β-LG＞α-LA，小分子质量葡聚糖会和各个蛋白质分子发生糖基化反应（Jiménez-Castaño et al.，2007）。

葡聚糖与乳清蛋白质的美拉德反应对乳清蛋白质的成胶性能有很大影响。将葡聚糖（15~25 kDa）与乳清蛋白质在 60℃、相对湿度 63%的条件下加热 2~9 天，可防止共轭凝胶在 80%形变的情况下发生断裂（Spotti et al.，2013），与单独的乳清蛋白质相比，25℃时乳清蛋白质/葡聚糖糖基化产物的凝胶时间、凝胶温度和储能模量都增加了（Spotti et al.，2014a，b）。但在 60℃加热 7 天制备的质量比为 1∶1 的乳清蛋白质/葡聚糖（150 kDa）产物，在 25℃时的储能模量却低于乳清蛋白质凝胶（Sun et al.，2012）。

2）果胶

对于阴离子多糖，乳清蛋白质与多糖的相互作用可涉及共价偶联和静电作用。果胶是一种线形多糖，由部分被甲氧基酯化的聚半乳糖醛酸亚基构成。为了避免形成离子配合物，通常会将果胶和 WPC 在干热条件下反应。经干热法（pH 7，60℃，5~15 天）制备的乳清蛋白质/果胶复合物在 pH 4.6 时溶解度、乳化活性提高，凝胶性和发泡性优于单独的 WPC，在复合物浓度 0.10%下，乳化活性为 98.56%，在 0.25%浓度下，乳化稳定性为 99.65%（Mishra et al.，2001）。多糖的"毛发"结构可增强空间排斥力，增加乳液的稳定性（Wooster and Augustin，2007），这与介质的 pH 有关。在 pH 7、60℃的条件下加热 14 天形成的乳清蛋白质/果胶复合物在 pH 5.5 时乳化性能得到了显著提高,但在 pH 4 时则会降低(Neirynck et al., 2004)。

2. 乳清蛋白质和糖类的湿热法美拉德反应

湿热法是美拉德反应的常用加工方法,尽管其褐变强度是不可控的。在高含水量下,水溶性反应物被稀释,反应速率会有所降低(Nasirpour et al., 2006)。已有研究利用湿热处理乳清蛋白质与碳水化合物形成复合物(Qi et al., 2017)。湿热可以有效地控制反应程度,但高温往往会引起蛋白质的聚集(Perusko et al., 2015)和其他反应。

在美拉德反应的初始阶段,乳清蛋白质和葡聚糖优化的反应条件为:10% WPI 与 30%葡聚糖,在 pH 6.5、60℃条件下加热 24 h,该条件下反应产物稳定(Zhu et al., 2008)。该温度(60℃)低于乳清蛋白质的变性温度(β-LG 78℃、α-LA 62℃、BSA 64℃)(Bryant and McClements, 1998)。将木糖和葡萄糖分别与乳清蛋白质在 50℃下加热 7 天,得到的美拉德反应产物具有较高的抗氧化活性。结构分析表明,美拉德反应改变了乳清蛋白质的酰胺Ⅰ带、酰胺Ⅱ带和酰胺Ⅲ带。圆二色谱分析表明,美拉德反应后β折叠、β转角和无规卷曲增加,而α螺旋则减少了(Wang et al., 2013)。

在高于变性温度时,BSA 更倾向于在溶液中与蔗糖相互作用,相对于其天然状态来说,该球状体的热稳定性得到了增强(Baier and McClements, 2001)。乳清蛋白质与麦芽糊精或玉米糖浆在 90℃、pH 8.2 下加热 24 h,可以提高乳清蛋白质在 pH 4.5 时的溶解度。在加入 NaCl(50 mmol/L)后,85℃加热 3 min,也会提高乳清蛋白质稳定性(Mulcahy et al., 2016)。美拉德反应程度与 BSA 凝胶的强度有关(Hill et al., 1992)。乳糖阻碍了 WPI 的变性,增加了凝胶时间和温度,降低了凝胶的断裂模量,但核糖有利于美拉德反应以及蛋白质的共价交联,同时也增加了凝胶的断裂模量,在含有核糖的凝胶中由美拉德反应引起的 pH 下降是在蛋白质变性和凝胶后发生的,对凝胶过程影响不大(Rich and Foegeding, 2000)。

4.4.2 乳清蛋白质和多糖的相互作用

由美拉德反应形成的生物聚合物分子会永久地连接在一起。非共价结合的复合物结合作用相对较弱,反应是可逆的。乳清蛋白质和多糖的相互作用在食品加工中是常见的现象,具有重要的意义。乳清蛋白质与多糖的相互作用取决于电荷密度、溶剂 pH、离子强度和聚合物结构(Doublier et al., 2000)。水中蛋白质和多糖不同类型的相互作用如图 4.2 所示。从热力学的观点来看,蛋白质和多糖在水溶液中可以是相容的或不相容的。在热力学不相容的条件下,得到了两相体系,其中两种不同的分子主要处于不同的相中。在热力学相容的条件下,可以得到两种类型的溶液：①均相稳定的体系，分子共存于同一相中；②两相体系，两种分子相互作用，本质上处于同一浓缩相，称为复合作用或复合凝聚（Laneuville et al., 2000）。复合凝聚是由两个带相反电荷的胶体相互作用引起的凝聚。

相分离可分为两类：①热力学不相容或互斥相分离；②亲和相分离（复合凝聚）（Doublier et al., 2000）。

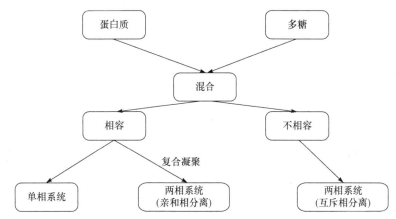

图 4.2 水介质中蛋白质和多糖之间不同类型的相互作用
资料修改自 Tolstoguzov（1991）

1. 乳清蛋白质与中性多糖的相互作用

1）菊粉

菊粉主要从菊苣根中提取，是经过纯化和结晶步骤得到的低聚体和末端有β-D-果糖或α-D-葡萄糖单元的链状聚合物的混合物。这些链状聚合物由 2~60 个不等的单元组成，平均聚合度约为 12。菊粉能提高蛋白质悬液的起泡稳定性，降低乳化活性和乳化稳定性（Herceg et al., 2007）。用菊粉代替乳糖可以提高乳清蛋白质的变性程度（Tobin et al., 2010），7%以上的乳清蛋白质，显著提高了菊粉凝胶（20%和 35%）的储能模量和损耗模量，并形成了更牢固的网络（Glibowski, 2009）。α-LA（1.4%）对凝胶性能无显著影响，β-LG（4.2%）显著提高菊粉凝胶（20%）的储能模量（Glibowski, 2010）。

2）淀粉

食品中的淀粉通常与其他大分子结合在一起，存在于谷物、水果、蔬菜和肉类等食物中，或作为增稠剂、稳定剂和胶凝剂添加到食品中。对乳清蛋白质和交联糯玉米淀粉的凝胶特性进行研究，差示扫描量热法（DSC）热扫描结果表明，这两种材料表现出独立的热跃迁，两种材料间没有相互作用（Fitzsimons et al., 2008a）。在用热诱导法制备 WPI（10%）凝胶时，用木薯淀粉替代少量蛋白质（0%~40%），力学性能得到了改善（Aguilera and Rojas, 1996），当木薯淀粉颗粒体积分数较小时（如小于颗粒的最大填充量），淀粉膨胀会使 WPI 凝胶强度增加（Aguilera and Rojas, 1997）。乳清蛋白质与小麦粉的相互作用可降低乳清蛋白质变性温度和 WPC 凝胶的蛋白质溶解性。不同 pH 下，小麦粉对 WPC 凝胶性能的影响也不同。在酸性条件下（pH 3.75），WPC 凝胶的松弛时间和黏结性随着小麦粉的加入而增加；中性条件下，松弛时间和黏结性则随之降低（Yamul and Lupano, 2005）。

2. 乳清蛋白质与阴离子多糖的相互作用

静电相互作用在蛋白质-阴离子多糖复合物的形成中起关键作用，pH、离子强度以及蛋白质与聚合电解质比例也有很大影响（Burova et al., 2007）。

1）果胶

果胶是一种由半乳糖醛酸单元通过 α-1,4 连接而成的线形多糖，中间穿插着由 α-1,2 连接的鼠李糖单元，其中半乳糖醛酸的羧基可以被甲基酯化。甲氧基化的程度是非常重要的，因为它影响果胶的行为。乳清蛋白质与果胶的相互作用取决于 pH、离子强度、浓度等因素。乳清蛋白质与果胶通过带负电荷的羧基与带正电荷的氨基酸残基之间的静电吸引产生相互作用。当 pH 低于乳清蛋白质的等电点时，乳清蛋白质带正电荷，而果胶带负电荷，产生络合作用。远离乳清蛋白质等电点时，带负电荷的果胶可以通过静电吸引与乳清蛋白质带正电的部分相互作用。

低甲氧基果胶与乳清蛋白质（0.45%）混合物在酸性条件下搅拌可形成聚集物，盐酸还可使更高浓度的乳清蛋白质/热处理乳清蛋白质（6.3%）混合物形成凝胶。果胶中和了热处理乳清蛋白质的正电荷，在 pH 1.5～3.0 下形成凝胶，且能改变热处理乳清蛋白质凝胶的性质。当 pH 为 3.5 和 4 时，热处理乳清蛋白质能形成凝胶，但热处理乳清蛋白质和果胶的混合物则无法形成凝胶（Li and Zhong, 2016）。将 WPI-低甲氧基果胶混合物（0.5%∶0.5%）调节至 pH 为 2 时，经加热-冷却步骤可形成凝胶。低 pH 抑制了低甲氧基果胶分子间的静电斥力，使 WPI 分子与低甲氧基果胶分子间的相互作用增强。低甲氧基果胶有助于构建作为支撑凝胶体系结构框架的网络，WPI 有助于网络结构的稳定（Wijaya et al., 2017）。pH 4～6 时，乳清蛋白质和果胶混匀后加热，促使蛋白质-多糖的相互作用超过蛋白质-蛋白质的聚集，zeta 电位的变化证明，在 pH 为 6 时，未加热的乳清蛋白质和果胶没有发生相互作用，加热后发生相互作用，在 pH 4 时，加热会使果胶与乳清蛋白质表面的阳离子发生强烈的相互作用（Krzeminski et al., 2014）。

热力学不相容是生物高分子混合物中常见的现象。如果体系中有凝胶形成，则在相分离和凝胶形成之间存在竞争。体系中的乳清蛋白质/果胶之间的斥力要强于乳清蛋白质/乳清蛋白质和果胶/果胶之间斥力的平均值，才能实现相分离。某些 pH 条件下，体系中乳清蛋白质/果胶之间静电斥力较低，乳清蛋白质更容易形成凝胶，并与果胶共同形成连续的网状结构。在 pH 为 6 时，低甲氧基果胶的存在使形成凝胶所需要的蛋白质浓度比正常情况下低。增加果胶的用量和钙离子浓度可使凝胶强度更大（Beaulieu et al., 2001）。将1%的果胶加入到8%的乳清蛋白质中，pH 8条件下80℃加热30 min，混合物形成凝胶，这两种聚合物都带有很高的负电荷，在蛋白质形成凝胶之前发生相分离，这些混合凝胶的结构主要是球形聚集体（Turgeon and Beaulieu, 2001）。

混合生物高分子的静电相互作用会引起混合体系流变性能的变化。蛋白质/多糖的酸凝胶是蛋白质凝胶和蛋白质聚集体与多糖分子之间的相分离相互竞争的结果（De Jong et al., 2009）。阴离子聚合电解质，如果胶，可以添加到体系中以促进预热的乳清蛋白质形成酸凝胶（Li and Zhong, 2016）。乳清蛋白质与果胶共同加热会减少相分离（Zhang and Vardhanabhuti, 2014）。pH为6~6.4时，果胶与3%乳清蛋白质的质量比0~0.2时的研究结果表明，通过与多糖形成可溶性复合物，蛋白质的热稳定性得到了增强，这是由于两种生物高分子之间的静电作用，多糖具有稳定蛋白质二级结构的能力，进而改变了蛋白质的热聚集行为（Zhang et al., 2012）。pH以及果胶的电荷密度极大地影响着蛋白质和果胶之间的相互作用和凝胶强度，在酸凝胶体系中，在pH为7.0的条件下加热得到的复合物凝胶，其强度显著提高，而在pH为6.5或6.2时，则不显著。该复合物具有较高的持水性和精细的结构（Zhang et al., 2014）。对于电荷密度高的果胶，复合凝胶具有更高的凝胶强度和更好的持水性。乳清蛋白质/果胶复合物具有平滑的网状结构（Zhang et al., 2014）。在pH为7时，果胶在凝胶形成前已经与蛋白质结合，并埋藏于最终凝胶的蛋白质网络中（Zhang et al., 2014）。

2）黄原胶

黄原胶（XG）是由野油菜黄单胞菌产生的一种阴离子细菌多糖，分子质量超过10^6 Da，脆弱的结构使其凝胶具有剪切稀释的特性，因此黄原胶可用于控制黏度（Panaras et al., 2011）。黄原胶分子具有有序的螺旋形构象，由一条纤维素的主链构成，带负电荷的三糖侧链沿主链排列（Laneuville et al., 2000）。酸性条件下，质子化的乳清蛋白质通过静电吸引与黄原胶的羧酸基团相互作用。当乳清蛋白质与黄原胶的比例为20:1、总固形物含量为1%、pH为5.4时，可以形成纤维状复合物（Laneuville et al., 2000）。在此条件下，黄原胶呈纤维状，乳清蛋白质可沿黄原胶聚集体形成纤维状复合物。在络合前，微射流处理黄原胶可产生颗粒状乳清蛋白质-黄原胶络合物，作为蛋糕奶油糖霜和夹心饼干馅的脂肪替代物（Laneuville et al., 2005）。在一定的pH和组分条件下，乳清蛋白质-黄原胶复合物能改善发泡性能（Xie and Hettiarachchy, 1998）。WPI（4%~10%）与黄原胶（0.5%）的混合溶液形成透明体系，表现出新的物理化学性质，如流变性能、表面特性、表面电荷密度（zeta电位）、表面疏水性以及扩散行为（Benichou et al., 2007）。

在混合体系中，生物高分子组分在胶体颗粒表面形成固定层（如胶体蛋白质聚集体或蛋白质稳定的粗乳液），三元溶液中的不相容性、混溶性和凝聚性转变为不吸附、弱吸附和强吸附，并受到 pH、离子强度和温度的影响（Syrbe et al.,1998）。黄原胶是一种不吸附的多糖，由于排斥絮凝机理，液滴的聚集可能发生在乳清蛋白质稳定的乳液中（Panaras et al., 2011）。中性 pH 条件下，黄原胶通过排斥机制使乳清蛋白质溶液中的预变性/聚集蛋白发生絮凝（Hemar et al., 2001），盐的存在会使黄原胶与热变性乳清蛋白质产生相分离，形成嵌在乳清蛋白质凝胶中的黄原胶富集区，导致凝胶的不透明度和硬度显著增加（Bryant and McClements, 2000）。

当混合溶液中的乳清蛋白质与黄原胶浓度足够高时，乳清蛋白质的凝胶特性发生改变。黄原胶同时具有协同和拮抗作用，这取决于 pH、盐的添加量、加热速率以及乳清蛋白质的变性程度。Sanchez 等（2010）发现在 pH≥7 时乳清蛋白质与黄原胶具有协同作用，而在 pH≤6.5 时具有拮抗作用。但在 pH 为 6.5 和 6.0 时，添加黄原胶（0.01%~0.06%）会导致相分离，乳清蛋白质更集中，凝胶强度更大，在 pH 高于乳清蛋白质等电点时加热，可以观察到乳清蛋白质与黄原胶之间的互斥相分离，这种混合凝胶是含有黄原胶的蛋白质连续相。在 pH 为 5.5，黄原胶添加量为 0.01%~0.06%时，均可观察到拮抗作用。在接近乳清蛋白质等电点的 pH 下，蛋白结构上的正电区域分布不均匀，使其与黄原胶发生络合，降低了蛋白凝胶团簇和/或聚集体之间的连接，从而使蛋白凝胶强度变弱（Bertrand and Turgeon, 2007）。

乳清蛋白质凝胶制备过程中相分离与凝胶形成的竞争关系受升温速率等变量的控制。Li 等（2006）研究了五种加热速率（0.1℃/min、1℃/min、5℃/min、10℃/min、20℃/min）对乳清蛋白质和黄原胶混合物凝胶的影响，发现 WPI 凝胶温度随着加热速率降低和黄原胶的加入而降低。当黄原胶含量为 0%~0.2%时，WPI 凝胶的断裂应力随升温速率减小而减小；当黄原胶含量为 0.5%和 1%时，断裂应力随升温速率增大而增大。

3）卡拉胶

卡拉胶（CG）是一种水溶性的、规则的线形多糖，来源于海洋中的红藻。卡拉胶是一种阴离子硫酸多糖。根据其一级结构的不同，卡拉胶被分为几种不同的类型（Burova et al., 2007）。卡拉胶的硫酸基团能与乳清蛋白质的质子化氨基相互作用。用于食品工业的卡拉胶主要有三种类型：阿欧塔型（ι, iota）、卡帕型（κ, kappa）、拉姆达型（λ, lambda）。阿欧塔型和卡帕型是凝胶型，当溶液冷却时会形成凝胶，再加热会恢复到溶液状态。凝胶形成过程中的主要变化是从无序的卷曲到同轴双螺旋的转变（Harrington et al., 2009）。拉姆达型卡拉胶没有明显的凝胶特性（Weinbreck et al., 2004）。

κ-卡拉胶作为稳定剂被广泛用于乳制品中。pH 接近 4.5 是乳清蛋白质和κ-卡拉胶通过静电引力形成络合物的最佳条件。WPI∶κ-卡拉胶为 12∶1 时，形成的络合物光密度（OD）值最高。NaCl 的加入破坏了 WPI-κ-卡拉胶的络合作用（Stone and Nickerson，2012）。WPI-κ-卡拉胶混合物的热处理导致蛋白质形成凝胶和相分离。当κ-卡拉胶与 WPI 聚集体混合时，可观察到排空作用引起的相分离（Gaaloul et al.，2010）。

κ-卡拉胶的加入对β-LG 变性温度（Capron et al.，1999）、乳清蛋白质变性速率（Tziboula and Horne，1999）以及加热过程中β-LG 和乳清蛋白质的天然蛋白质损耗速率（Croguennoc et al.，2001a，b；De la Fuente et al.，2004）没有影响。然而，κ-卡拉胶的存在增加了"可溶性"乳清蛋白质聚集体的多分散性，减小了 75℃时聚集体的大小（De la Fuente et al.，2004）。其主要原因是κ-卡拉胶的排斥-絮凝作用，而不是相分离（De la Fuente et al.，2004）。在较高的温度下，聚集机制似乎发生了变化。在 0.1% κ-卡拉胶、pH 7.0 条件下，85℃加热浓度为 0.5%的 WPI 时形成了更高分子量的中间体。有κ-卡拉胶存在时，90℃加热 WPI 的聚集程度要大得多（Flett and Corredig，2009）。虽然不发生可见的相分离，但多糖和乳清蛋白质表现出热力学不相容（Flett and Corredig，2009）。

不同 pH 条件下β-LG/κ-卡拉胶的性质不同，pH 7、45℃下制备的混合物在冷却至 20℃过程中，会形成包含天然β-LG 的κ-卡拉胶凝胶网络。其天然状态下，特别是在高浓度下，蛋白质可能削弱了多糖凝胶网络（Eleya and Turgeon，2000 b）。pH 5～7 条件下将β-LG/κ-卡拉胶加热到 90℃，由于相分离，β-LG 与κ-卡拉胶在凝胶形成和熔化时的性质可以在混合体系中被识别出来。pH 4.0 时，由于β-LG 和κ-卡拉胶之间的静电吸引力，混合凝胶表现出与纯蛋白凝胶相似的特殊性质，即在蛋白质的等电点下，形成一个连续的网络结构（Eleya and Turgeon，2000a）。

在中性 pH 条件下，乳清蛋白质没有改变κ-卡拉胶（2.5%）的黏度和凝胶强度，乳清蛋白质和κ-卡拉胶的混合物没有发生相分离。虽然聚集的乳清蛋白质和κ-卡拉胶存在排斥絮凝作用，但聚集的乳清蛋白质数量太少，不能对κ-卡拉胶的流变性能产生影响。天然乳清蛋白质与κ-卡拉胶在中性 pH 下的相互作用并不显著（Hemar et al.，2002）。pH 7 时，将浓度为 5%的 WPI 与不同浓度的κ-卡拉胶（0%、0.2%、0.3%、0.5%和 0.6%）混合，当κ-卡拉胶浓度大于 0.3%时，混合物黏度急剧上升，且与分离条件有关。凝胶和相分离之间的竞争取决于剪切处理是在凝胶形成时应用还是在形成之后应用（Gaaloul et al.，2009）。

λ-卡拉胶带电电荷很多，每个重复单位含有三个硫酸基团。Weinbreck 等（2004）研究了在不同条件下，乳清蛋白质与 λ-卡拉胶之间的静电相互作用，可溶或不可溶复合物的形成取决于多种参数，如 pH、离子强度和聚合物比例。在钙离子存在时，可以通过钙离子结合形成 pH 为中性的络合物。如果在体系中加入 45 mmol/L 的 NaCl，则不溶性复合物的形成会增强，复合物会在蛋白质的等电点附近以及低于等电点的条件下形成（Weinbreck et al.，2004）。Stone 和 Nickerson（2012）发现乳清蛋白质-λ-卡拉胶（0.25%）的初始缔合开始于 pH 大于等电点时，这是由于在蛋白质表面存在正电荷。与各自单独的体系相比，乳清蛋白质/λ-卡拉胶复合物的乳液稳定性得到增强。

λ-卡拉胶的存在促进了低浓度（8%）MPC 的凝胶作用（Spahn et al.，2008）。λ-卡拉胶与乳清蛋白质的混合溶液具有剪切稀释的特性，表观黏度高于单独的两种溶液（Lizarraga et al.，2006）。

4.5 乳清蛋白质和食品中其他成分的相互作用

4.5.1 明胶

动物的皮肤和骨骼富含胶原蛋白，它是结缔组织的结构蛋白。明胶是由胶原蛋白经部分水解产生的，是不同大小的胶原蛋白碎片的多分散混合物（Hernàndez-Balada et al.，2009）。在明胶分子中有三组主要的氨基酸，它们是甘氨酸、丙氨酸和脯氨酸/羟脯氨酸。明胶含有大量的羟基、羧基和氨基，具有独特的成胶性能（Djagny et al.，2001）。

在大多数情况下，两种或两种以上的生物聚合物混合会导致相分离，可以是亲和的（两种聚合物均富集在第一相中，溶剂处于第二相中），也可以是互斥的（两种聚合物各自富集在一相中）（Dickinson et al.，2003；Tolstoguzov，1995）。乳清蛋白质和明胶是互斥相分离。乳清蛋白质的凝胶作用不受明胶的影响。由于混合物的熵很低，因此不发生相互作用。不同的混合凝胶有不同的相转变点（Walkenström and Hermansson，1994）。pH 远离等电点时，乳清蛋白质可以形成链状网络结构。pH 为 7.5 和 8.5 时，明胶和乳清蛋白质形成独立的相分离凝胶（Walkenström and Hermansson，1996）。在 10% WPC 和 3%的明胶中可以观察到凝胶性质转变。这种转变是由明胶决定，并被解释为明胶的连续化（Walkenström and Hermansson，1996）。在高压处理下，乳清蛋白质和明胶混合凝胶在 pH 7 时也表现出相分离行为。高压处理后的混合凝胶与热处理后的混合凝胶相比，具有更高程度的明胶连续性，其密集的网状结构也使乳清蛋白质和明胶难以鉴别（Walkenström and Hermansson，1997a）。在 pH 7.5 的条件下，结合加热和压力，混合凝胶由明胶和乳清蛋白质均匀分布的单一聚合网络组成。在混合网络中，明胶和乳清蛋白质无法区分（Walkenström and Hermansson，1997b）。乳清蛋白质（5%～20%）与 5%明胶在 5℃或 30℃、600 MPa 高压处理 15 min 后形成混合凝胶，共聚焦图像证实混合凝胶中明胶是连续相，而乳清蛋白质在加压混合体系中以不连续的聚集体形式存在（Devi et al.，2014）。明胶和聚合乳清蛋白质之间的排斥作用也得到了 Fitzsimons 等（2008b）的证实，他们发现在凝胶形成的初期阶段，明胶分子的结合导致相分离，形成了一个连续的明胶凝胶基质，聚合乳清蛋白质以小液滴状分散在凝胶中。虽然 Walkenström 和 Hermansson 很好地解释了乳清蛋白质和明胶之间的相分离，但 Sarbon 等（2015）分离了鸡皮明胶，并研究了鸡皮明胶对乳清蛋白质流变性和热性能的影响。他们发现，由于协同作用，明胶增加了乳清蛋白质的弹性模量，鸡皮明胶的加入也提高了乳清蛋白质的转变温度。

4.5.2 卵磷脂

卵磷脂是一种两性离子表面活性剂，由磷脂混合物组成，磷脂是细胞膜的重要成分。卵磷脂是由磷脂酰胆碱（PC）、磷脂酰乙醇胺（PE）、磷脂酰丝氨酸（PS）、磷脂酰肌醇（PI）和磷脂酰甘油（PG）等磷脂衍生物组成的混合物。卵磷脂与乳清蛋白质之间的相互作用主要涉及静电相互作用和疏水相互作用，原因是磷脂具有负电荷和烷基链。

在生理条件下（pH 7），β-LG 可能参与一些小型疏水性和两亲性化合物活动，如磷脂的运输。β-LG 与二肉豆蔻酰基磷脂酰甘油酯（dimyristoylphosphatidylglycerol）相互作用导致β-LG 二级结构变化，α螺旋含量增加，蛋白质三级结构展开。蛋白质的构象变化首先是由静电作用引起的，其次是由带负电荷的蛋白质与带负电荷的磷脂之间的疏水作用引起的（Liu et al.，2006）。在 pH 为 3.7 时，β-LG 与磷脂酰胆碱复合物的形成也使β-LG 的α螺旋含量增加。在 pH 为 7.2 时，磷脂酰胆碱和β-LG 形成聚集体（Brown et al.，1983）。在胃液（酸性）条件下，部分展开的α-LA 可穿透磷脂酰胆碱囊泡，与磷脂脂肪酸链产生疏水的相互作用，从而通过减少蛋白质与胃蛋白酶的接触，来减缓胃内的消化过程（Moreno et al.，2005）。α-LA 在酸性条件下被嵌入到磷脂酰丝氨酸/磷脂酰乙醇胺的囊泡双层膜中，保护其免受胰蛋白酶和胰凝乳蛋白酶的蛋白质水解作用（Kim and Kim，1986）。磷脂酰胆碱不影响β-LG 对胃液蛋白酶酶解的抗性，它使蛋白质在十二指肠环境下不被降解，脂质结合到β-LG 的一个二级脂肪酸结合位点，从而在空间上阻断了蛋白酶的作用（Mandalari et al.，2009）。

表面活性剂是较低浓度的单体，而胶束形成于较高浓度或高于临界胶束浓度。大豆卵磷脂在临界胶束浓度以下或以上均对乳清蛋白质原纤维有不同程度的影响。在临界胶束浓度以下，卵磷脂没有改变乳清蛋白质的线形结构和长原纤维。而在酸性条件下，乳清蛋白质部分展开，卵磷脂通过疏水作用与乳清蛋白质相互作用，导致α螺旋结构增加，β折叠结构减少。这种离子型表面活性剂通过电荷相互作用，引起蛋白质的初始变性。在临界胶束浓度以上、卵磷脂存在的条件下可观察到乳清蛋白质的聚集（Mantovani et al.，2016）。

蛋白质结合溶血磷脂酰胆碱（LPC）极大地减少了热诱导的蛋白质分子间的

相互作用，从而阻止了热诱导的蛋白质聚集（Le et al.，2011）。由于疏水相互作用，磷脂可以改变乳清蛋白质的二级结构（如在液晶状态下使用阴离子磷脂时）（Kasinos et al.，2013）。磷脂酰胆碱的加入可以形成卵磷脂-蛋白质络合物，提高凝胶强度（Brown，1984）。卵磷脂对乳清蛋白质热诱导凝胶的影响取决于凝胶类型。通过改变氯化钠的浓度，热诱导 WPI 形成细链、混合和微粒状网络结构。对于细链或混合的网状结构（低离子浓度），蛋黄卵磷脂（约含有 60%的磷脂酰胆碱和 20%的磷脂酰乙醇胺）在冷却过程中加速了凝胶化速率并增强了网状结构的弹性（Ikeda and Foegeding，1999）。高离子浓度会屏蔽乳清蛋白质的有效电荷，影响卵磷脂的亲疏水平衡，这可能是卵磷脂对微粒状网络结构凝胶化速率影响不大的原因（Ikeda and Foegeding，1999）。

4.6 总　　结

乳清蛋白质是一类具有高级结构的球状蛋白质，包括一级、二级、三级甚至四级结构。本章讨论了酶法处理和物理处理（加热、辐射和超声波）对乳清蛋白质构象和理化性质的影响。本章的另一个主题是乳清蛋白质和其他生物高分子之间的相互作用。乳清蛋白质分子含有游离巯基、带电的氨基酸残基、ε-氨基和疏水区域，可与模型系统或食品中的其他蛋白质、碳水化合物和脂类发生相互作用。乳清蛋白质与其他聚合物的相互作用有助于其功能上的改变，如溶解度、乳化容量和凝胶性质。

参 考 文 献

Aguilera, J. and Rojas, E. (1996). Rheological, thermal and microstructural properties of whey proteincassava starch gels. *Journal of Food Science* **61** (5): 962-966.

Aguilera, J.M. and Rojas, G.V. (1997). Determination of kinetics of gelation of whey protein and cassava starch by oscillatory rheometry. *Food Research International* **30** (5): 349-357.

Akhtar, M. and Dickinson, E. (2007). Whey protein-maltodextrin conjugates as emulsifying agents: an alternative to gum arabic. *Food Hydrocolloids*, **21** (4): 607-616.

Alting, A.C., Hamer, R.J., de Kruif, C.G. et al. (2000). Formation of disulfide bonds in acid-induced gels of preheated whey protein isolate. *Journal of Agricultural and Food Chemistry* **48** (10): 5001-5007.

Alting, A.C., Hamer, R.J., de Kruif, C.G. et al. (2003). Number of thiol groups rather than the size of the aggregates determines the hardness of cold set whey protein gels. *Food Hydrocolloids*, **17** (4): 469-479.

Anema, S.G. (2000). Effect of milk concentration on the irreversible thermal denaturation and disulfide aggregation of β-lactoglobulin. *Journal of Agricultural and Food Chemistry*, **48** (9):

4168-4175.

Ashokkumar, M. (2015). Applications of ultrasound in food and bioprocessing. *Ultrasonics Sonochemistry* **25**: 17-23.

Baier, S. and McClements, D.J. (2001). Impact of preferential interactions on thermal stability and gelation of bovine serum albumin in aqueous sucrose solutions. *Journal of Agricultural and Food Chemistry*, **49** (5): 2600-2608.

Beaulieu, M., Turgeon, S.L., and Doublier, J.L. (2001). Rheology, texture and microstructure of whey proteins/low methoxyl pectins mixed gels with added calcium. *International Dairy Journal*, **11** (11-12): 961-967.

Benichou, A., Aserin, A., Lutz, R. et al. (2007). Formation and characterization of amphiphilic conjugates of whey protein isolate (WPI)/xanthan to improve surface activity. *Food Hydrocolloids*, **21** (3): 379-391.

Bertrand, M.E. and Turgeon, S.L. (2007). Improved gelling properties of whey protein isolate by addition of xanthan gum. *Food Hydrocolloids* **21** (2): 159-166.

Bhattacharyya, J. and Das, K.P. (1999). Molecular chaperone-like properties of an unfolded protein, αscasein. *Journal of Biological Chemistry* **274** (22): 15505-15509.

Boye, J. and Alli, I. (2000). Thermal denaturation of mixtures of α-lactalbumin and β-lactoglobulin: a differential scanning calorimetric study. *Food Research International* **33** (8): 673-682.

Boye, J., Ma, C., and Harwalkar, V. (1997a). Thermal denaturation and coagulation of proteins. In: *Food Science and Technology*, 25-56. NY: Marcel Dekker.

Boye, J.I., Ma, C.Y., Ismail, A. et al. (1997b). Molecular and microstructural studies of thermal denaturation and gelation of beta-lactoglobulins A and B. *Journal of Agricultural and Food Chemistry* **45** (45): 1608-1618.

Brown, E.M. (1984). Interactions of β-lactoglobulin and α-lactalbumin with lipids: a review. *Journal of Dairy Science* **67** (4): 713-722.

Brown, E.M., Carroll, R.J., Pfeffer, P.E. et al. (1983). Complex formation in sonicated mixtures of β-lactoglobulin and phosphatidylcholine. *Lipids* **18** (2): 111-118.

Bryant, C.M. and McClements, D.J. (1998). Molecular basis of protein functionality with special consideration of cold-set gels derived from heat-denatured whey. *Trends in Food Science and Technology* **9** (4): 143-151.

Bryant, C.M. and Mcclements, D.J. (2000). Influence of xanthan gum on physical characteristics of heat-denatured whey protein solutions and gels. *Food Hydrocolloids* **14** (4): 383-390.

Burova, T.V., Grinberg, N.V., Grinberg, V.Y. et al. (2007). Conformational changes in ι-and κ-carrageenans induced by complex formation with bovine β-casein. *Biomacromolecules* **8** (2): 368-375.

Calvo, M.M., Leaver, J., and Banks, J.M. (1993). Influence of other whey proteins on the heat-induced aggregation of α-lactalbumin. *International Dairy Journal* **3** (8): 719-727.

Capron, I., Nicolai, T., and Durand, D. (1999). Heat induced aggregation and gelation of β-lactoglobulin in the presence of κ-carrageenan. *Food Hydrocolloids* **13** (1): 1-5.

Chandrapala, J., Zisu, B., Palmer, M. et al. (2011). Effects of ultrasound on the thermal and structural characteristics of proteins in reconstituted whey protein concentrate. *Ultrasonics Sonochemistry* **18** (5): 951-957.

Chandrapala, J., Zisu, B., Kentish, S. et al. (2012). The effects of high-intensity ultrasound on the structural and functional properties of α-lactalbumin, β-lactoglobulin and their mixtures. *Food Research International* **48** (2): 940-943.

Chawla, S., Chander, R., and Sharma, A. (2009). Antioxidant properties of Maillard reaction products obtained by gamma-irradiation of whey proteins. *Food Chemistry* **116** (1): 122-128.

Cieśla, K., Salmieri, S., Lacroix, M. et al. (2004). Gamma irradiation influence on physical properties of milk proteins. *Radiation Physics and Chemistry* **71** (1-2): 95-99.

Cieśla, K., Salmieri, S., and Lacroix, M. (2006). Modification of the properties of milk protein films by gamma radiation and polysaccharide addition. *Journal of the Science of Food and Agriculture* **86** (6): 908-914.

Croguennoc, P., Durand, D., Nicolai, T. et al. (2001a). Phase separation and association of globular protein aggregates in the presence of polysaccharides: 1. Mixtures of preheated β-lactoglobulin and κ- carrageenan at room temperature. *Langmuir* **17** (14): 4372-4379.

Croguennoc, P., Nicolai, T., Durand, D. et al. (2001b). Phase separation and association of globular protein aggregates in the presence of polysaccharides: 2. Heated mixtures of native β-lactoglobulin and κ-carrageenan. *Langmuir* **17** (14): 4380-4385.

Dai, Q., Zhu, X., Abbas, S. et al. (2015). Stable nanoparticles prepared by heating electrostatic complexes of whey protein isolate-dextran conjugate and chondroitin sulfate. *Journal of Agricultural and Food Chemistry* **63** (16): 4179-4189.

Dalgleish, D.G. (2011). On the structural models of bovine casein micelles-review and possible improvements. *Soft Matter* **7** (6): 2265-2272.

Dalgleish, D.G. and Corredig, M. (2012). The structure of the casein micelle of milk and its changes during processing. *Annual Review of Food Science and Technology* **3**: 449-467.

De Jong, S., Klok, H.J., and Van de Velde, F. (2009). The mechanism behind microstructure formation in mixed whey protein-polysaccharide cold-set gels. *Food Hydrocolloids* **23** (3): 755-764.

De la Fuente, M., Hemar, Y., and Singh, H. (2004). Influence of κ-carrageenan on the aggregation behaviour of proteins in heated whey protein isolate solutions. *Food Chemistry* **86** (1): 1-9.

De Wit, J.N. and Swinkels, G.A.M. (1980). A differential scanning calorimetric study of the thermal denaturation of bovine β-lactoglobulin thermal behaviour at temperatures up to 100 ℃. *Biochimica et Biophysica Acta (BBA)-Protein Structure* **624** (1): 40-50.

DeJong, G.A.H. and Koppelman, S.J. (2002). Transglutaminase catalyzed reactions: impact on food applications. *Journal of Food Science* **67** (8): 2798-2806.

Devi, A.F., Buckow, R., Hemar, Y. et al. (2014). Modification of the structural and rheological properties of whey protein/gelatin mixtures through high pressure processing. *Food Chemistry* **156** (3): 243-249.

Dickinson, E., Evison, J., Owusu, R. et al. (2003). Gums and stabilisers for the food industry. *Proceedings of the 12th Gums and Stabilisers for the Food Industry-Designing Structure into Food*s, Wrexham, Wales (23-27 June 2003). Cambridge, UK: The Royal Society of Chemistry.

Djagny, K.B., Wang, Z., and Xu, S. (2001). Gelatin: a valuable protein for food and pharmaceutical industries. *Critical Reviews in Food Science and Nnutrition* **41** (6): 481-492.

Donato, L. and Guyomarc'h, F. (2009). Formation and properties of the whey protein/kappa-casein complexes in heated skim milk-a review. *Dairy Science and Technology* **89** (1): 3-29.

Donato, L., Guyomarc'h, F., Amiot, S. et al. (2007). Formation of whey protein/κ-casein complexes in heated milk: preferential reaction of whey protein with κ-casein in the casein micelles. *International Dairy Journal* **17** (10): 1161-1167.

Doublier, J.L., Garnier, C., Renard, D. et al. (2000). Protein-polysaccharide interactions. *Current Opinion in Colloid and Interface Science* **5** (3-4): 202-214.

Eissa, A.S., Satisha Bisram, A., and Khan, S.A. (2004). Polymerization and gelation of whey protein isolates at low pH using transglutaminase enzyme. *Journal of Agricultural and Food Chemistry* **52** (14): 4456-4464.

Eleya, M.O. and Turgeon, S. (2000a). The effects of pH on the rheology of β-lactoglobulin/κ-carrageenan mixed gels. *Food Hydrocolloids* **14** (3): 245-251.

Eleya, M.O. and Turgeon, S. (2000b). Rheology of κ-carrageenan and β-lactoglobulin mixed gels. *Food Hydrocolloids* **14** (1): 29-40.

Fitzsimons, S.M., Mulvihill, D.M., and Morris, E.R. (2008a). Co-gels of whey protein isolate with crosslinked waxy maize starch: analysis of solvent partition and phase structure by polymer blending laws. *Food Hydrocolloids* **22** (3): 468-484.

Fitzsimons, S.M., Mulvihill, D.M., and Morris, E.R. (2008b). Segregative interactions between gelatin and polymerised whey protein. *Food Hydrocolloids* **22** (3): 485-491.

Flett, K.L. and Corredig, M. (2009). Whey protein aggregate formation during heating in the presence of κ-carrageenan. *Food Chemistry* **115** (4): 1479-1485.

Frydenberg, R.P., Hammershoj, M., Andersen, U. et al. (2016). Protein denaturation of whey protein isolates (WPIs) induced by high intensity ultrasound during heat gelation. *Food Chemistry* **192**: 415-423.

Gaaloul, S., Corredig, M., and Turgeon, S.L. (2009). Rheological study of the effect of shearing process and κ-carrageenan concentration on the formation of whey protein microgels at pH 7. *Journal of Food Engineering* **95** (2): 254-263.

Gaaloul, S., Turgeon, S.L., and Corredig, M. (2010). Phase behavior of whey protein aggregates/κ-carrageenan mixtures: experiment and theory. *Food Biophysics* **5** (2): 103-113.

Galani, D. and Owusu Apenten, R.K. (2010). Heat-induced denaturation and aggregation of betalactoglobulin: kinetics of formation of hydrophobic and disulphide-linked aggregates. *International Journal of Food Science and Technology* **34** (5-6): 467-476.

Gauche, C., Barreto, P.L.M., and Bordignonluiz, M.T. (2010). Effect of thermal treatment on whey protein polymerization by transglutaminase: implications for functionality in processed dairy foods. *LWT-Food Science and Technology* **43** (2): 214-219.

Glibowski, P. (2009). Rheological properties and structure of inulin-whey protein gels. *International Dairy Journal* **19** (8): 443-449.

Glibowski, P. (2010). Effect of α-lactalbumin and β-lactoglobulin on inulin gelation.

Milchwissenschaft **65** (2): 127-131.

Gordon, L. and Pilosof, A.M. (2010). Application of high-intensity ultrasounds to control the size of whey proteins particles. *Food Biophysics* **5** (3): 203-210.

Gracia-Julia, A., Rene, M., Cortes-Munoz, M. et al. (2008). Effect of dynamic high pressure on whey protein aggregation: a comparison with the effect of continuous short-time thermal treatments. *Food Hydrocolloids* **22** (6): 1014-1032.

Gülseren, İ., Guzey, D., Bruce, B.D. et al. (2007). Structural and functional changes in ultrasonicated bovine serum albumin solutions. *Ultrasonics Sonochemistry* **14** (2): 173-183.

Guyomarc'h, F., Nono, M., Nicolai, T. et al. (2009). Heat-induced aggregation of whey proteins in the presence of κ-casein or sodium caseinate. *Food Hydrocolloids*, **23** (4): 1103-1110.

Guyomarc'h, F., Violleau, F., Surel, O. et al. (2010). Characterization of heat-induced changes in skim milk using asymmetrical flow field-flow fractionation coupled with multiangle laser light scattering. *Journal of Agricultural and Food Chemistry* **58** (24): 12592-12601.

Harrington, J., Foegeding, E., Mulvihill, D. et al. (2009). Segregative interactions and competitive binding of Ca^{2+} in gelling mixtures of whey protein isolate with Na^+ κ-carrageenan. *Food Hydrocolloids* **23** (2): 468-489.

Havea, P., Singh, H., and Creamer, L.K. (2001). Characterization of heat-induced aggregates of betalactoglobulin, alpha-lactalbumin and bovine serum albumin in a whey protein concentrate environment. *Journal of Dairy Research* **68** (3): 483-497.

Havea, P., Carr, A.J., and Creamer, L.K. (2004). The roles of disulphide and non-covalent bonding in the functional properties of heat-induced whey protein gels. *Journal of Dairy Research* **71** (3): 330-339.

Hemar, Y., Tamehana, M., Munro, P.A. et al. (2001). Viscosity, microstructure and phase behavior of aqueous mixtures of commercial milk protein products and xanthan gum. *Food Hydrocolloids* **15** (4-6): 565-574.

Hemar, Y., Hall, C., Munro, P. et al. (2002). Small and large deformation rheology and microstructure of κ-carrageenan gels containing commercial milk protein products. *International Dairy Journal* **12** (4): 371-381.

Herceg, Z., Režek, A., Lelas, V. et al. (2007). Effect of carbohydrates on the emulsifying, foaming and freezing properties of whey protein suspensions. *Journal of Food Engineering* **79** (1): 279-286.

Hernàndez-Balada, E., Taylor, M.M., Phillips, J.G. et al. (2009). Properties of biopolymers produced by transglutaminase treatment of whey protein isolate and gelatin. *Bioresource Technology* **100** (14): 3638-3643.

Hill, S., Mitchell, J., and Armstrong, H. (1992). The production of heat stable gels at low protein concentration by the use of the Maillard reaction. In: *Gums and Stabilizers for the Food Industry*, 479-485. Oxford, UK: Oxford University Press.

Hirano, R., Hirano, M., and Hatanaka, K. (1999). *Changes in Hardness of Acid Milk Gel by Addition of Hypothiocyanite Ion, Hydrogen Peroxide and their Effects on Sulfhydryls in Milk Proteins*. Food and Agriculture Organization of the United Nations.

Hoffmann, M.A. and van Mil, P.J. (1997). Heat-induced aggregation of β-lactoglobulin: role of the free thiol group and disulfide bonds. *Journal of Agricultural and Food Chemistry* **45** (8): 2942-2948.

Ikeda, S. and Foegeding, E. (1999). Dynamic viscoelastic properties of thermally induced whey protein isolate gels with added lecithin. *Food Hydrocolloids* **13** (3): 245-254.

Jambrak, A.R., Mason, T.J., Lelas, V. et al. (2014). Effect of ultrasound treatment on particle size and molecular weight of whey proteins. *Journal of Food Engineering* **121**: 15-23.

Jimenez-Castano, L., Villamiel, M., Martin-Alvarez, P.J. et al. (2005). Effect of the dry-heating conditions on the glycosylation of β-lactoglobulin with dextran through the Maillard reaction. *Food Hydrocolloids* **19** (5): 831-837.

Jimenez-Castano, L., Villamiel, M., and Lopez-Fandino, R. (2007). Glycosylation of individual whey proteins by Maillard reaction using dextran of different molecular mass. *Food Hydrocolloids* **21** (3): 433-443.

Kasinos, M., Sabatino, P., Vanloo, B. et al. (2013). Effect of phospholipid molecular structure on its interaction with whey proteins in aqueous solution. *Food Hydrocolloids* **32** (2): 312-321.

Kehoe, J. and Foegeding, E. (2010). Interaction between β-casein and whey proteins as a function of pH and salt concentration. *Journal of Agricultural and Food Chemistry* **59** (1): 349-355.

Kim, J. and Kim, H. (1986). Fusion of phospholipid vesicles induced by alpha-lactalbumin at acidic pH. *Biochemistry* **25** (24): 7867-7874.

Krzeminski, A., Prell, K.A., Weiss, J. et al. (2014). Environmental response of pectin-stabilized whey protein aggregates. *Food Hydrocolloids* **35**: 332-340.

Laneuville, S.I., Paquin, P., and Turgeon, S.L. (2000). Effect of preparation conditions on the characteristics of whey protein-xanthan gum complexes. *Food Hydrocolloids* **14** (4): 305-314.

Laneuville, S.I., Paquin, P., and Turgeon, S.L. (2005). Formula optimization of a low-fat food system containing whey protein isolate-xanthan gum complexes as fat replacer. *Journal of Food Science* **70** (8): S513-S519.

Le Tien, C., Letendre, M., Ispas-Szabo, P. et al. (2000). Development of biodegradable films from whey proteins by cross-linking and entrapment in cellulose. *Journal of Agricultural and Food Chemistry* **48** (11): 5566-5575.

Le, T.T., Sabatino, P., Heyman, B. et al. (2011). Improved heat stability by whey protein-surfactant interaction. *Food Hydrocolloids* **25** (4): 594-603.

Lee, J.W., Kim, J.H., Yook, H.S. et al. (2001). Effects of gamma radiation on the allergenic and antigenic properties of milk proteins. *Journal of Food Protection* **64** (2): 272-276.

Leeb, E., Haller, N., Kulozik, U. et al. (2017). Effect of pH on the reaction mechanism of thermal denaturation and aggregation of bovine β-lactoglobulin. *International Dairy Journal* **78**: 103-111.

Li, K. and Zhong, Q. (2016). Aggregation and gelation properties of preheated whey protein and pectin mixtures at pH 1.0-4.0. *Food Hydrocolloids* **60**: 11-20.

Li, J., Eleya, M.M.O., and Gunasekaran, S. (2006). Gelation of whey protein and xanthan mixture: effect of heating rate on rheological properties. *Food Hydrocolloids* **20** (5): 678-686.

Liu, G. and Zhong, Q. (2012). Glycation of whey protein to provide steric hindrance against thermal aggregation. *Journal of Agricultural and Food Chemistry* **60** (38): 9754-9762.

Liu, G. and Zhong, Q. (2013). Thermal aggregation properties of whey protein glycated with various saccharides. *Food Hydrocolloids* **32** (1): 87-96.

Liu, X., Shang, L., Jiang, X. et al. (2006). Conformational changes of β-lactoglobulin induced by anionic phospholipid. *Biophysical Chemistry* **121** (3): 218-223.

Lizarraga, M., Vicin, D.D.P., Gonzalez, R. et al. (2006). Rheological behaviour of whey protein concentrate and λ-carrageenan aqueous mixtures. *Food Hydrocolloids* **20** (5): 740-748.

Ma, S., Shi, C., Wang, C. et al. (2017). Effects of ultrasound treatment on physiochemical properties and antimicrobial activities of whey protein-totarol nanoparticles. *Journal of Food Protection* **80** (10): 1657-1665.

Ma, S., Wang, C., and Guo, M. (2018). Changes in structure and antioxidant activity of β-lactoglobulin by ultrasound and enzymatic treatment. *Ultrasonics Sonochemistry* **43**: 227-236.

Mandalari, G., Mackie, A.M., Rigby, N.M. et al. (2009). Physiological phosphatidylcholine protects bovine β-lactoglobulin from simulated gastrointestinal proteolysis. *Molecular Nutrition & Food Research* **53** (S1): 131-139.

Mantovani, R.A., Fattori, J., Michelon, M. et al. (2016). Formation and pH-stability of whey protein fibrils in the presence of lecithin. *Food Hydrocolloids* **60**: 88-298.

Matsudomi, N., Rector, D., and Kinsella, J. (1991). Gelation of bovine serum albumin and β-lactoglobulin; effects of pH, salts and thiol reagents. *Food Chemistry* **40** (1): 55-69.

Matsudomi, N., Oshita, T., Sasaki, E. et al. (1992). Enhanced heat-induced gelation of betalactoglobulin by alpha-lactalbumin. *Bioscience, Biotechnology, and Biochemistry* **56** (11): 1697-1700.

Mishra, S., Mann, B., and Joshi, V. (2001). Functional improvement of whey protein concentrate on interaction with pectin. *Food Hydrocolloids* **15** (1): 9-15.

Monahan, F.J., German, J.B., and Kinsella, J.E. (1995). Effect of pH and temperature on protein unfolding and thiol/disulfide interchange reactions during heat-induced gelation of whey proteins. *Journal of Agricultural and Food Chemistry* **43** (1): 46-52.

Morand, M., Guyomarc'h, F., Pezennec, S. et al. (2011). On how κ-casein affects the interactions between the heat-induced whey protein/κ-casein complexes and the casein micelles during the acid gelation of skim milk. *International Dairy Journal* **21** (9): 670-678.

Morand, M., Guyomarc'h, F., Legland, D. et al. (2012). Changing the isoelectric point of the heatinduced whey protein complexes affects the acid gelation of skim milk. *International Dairy Journal* **23** (1): 9-17.

Moreno, F.J., Mackie, A.R., and Mills, E.C. (2005). Phospholipid interactions protect the milk allergen α-lactalbumin from proteolysis during in vitro digestion. *Journal of Agricultural and Food Chemistry* **53** (25): 9810-9816.

Morgan, F., Molle, D., Henry, G. et al. (1999). Glycation of bovine β-lactoglobulin: effect on the protein structure. *International Journal of Food Science and Technology* **34** (5-6): 429-435.

Moro, A., Carlos Gatti, A., and Delorenzi, N. (2001). Hydrophobicity of whey protein concentrates measured by fluorescence quenching and its relation with surface functional properties. *Journal of Agricultural and Food Chemistry* **49** (10): 4784-4789.

Mulcahy, E.M., Mulvihill, D.M., and O'Mahony, J.A. (2016). Physicochemical properties of whey protein conjugated with starch hydrolysis products of different dextrose equivalent values. *International Dairy Journal* **53**: 20-28.

Nasirpour, A., Scher, J., and Desobry, S. (2006). Baby foods: formulations and interactions (a review). *Critical Reviews in Food Science and Nutrition* **46** (8): 665-681.

Neirynck, N., Van Der Meeren, P., Gorbe, S.B. et al. (2004). Improved emulsion stabilizing properties of whey protein isolate by conjugation with pectins. *Food Hydrocolloids* **18** (6): 949-957.

Oliveira, C.L.P., Hoz, L., Silva, J.C. et al. (2007). Effects of gamma radiation on is β-lactoglobulin: oligomerization and aggregation. *Biopolymers* **85** (3): 284-294.

Panaras, G., Moatsou, G., Yanniotis, S. et al. (2011). The influence of functional properties of different whey protein concentrates on the rheological and emulsification capacity of blends with xanthan gum. *Carbohydrate Polymers* **86** (2): 433-440.

Paulsson, M. and Dejmek, P. (1990). Thermal denaturation of whey proteins in mixtures with caseins studied by differential scanning calorimetry. *Journal of Dairy Science* **73** (3): 590-600.

Perusko, M., Al-Hanish, A., Velickovic, T.C. et al. (2015). Macromolecular crowding conditions enhance glycation and oxidation of whey proteins in ultrasound-induced Maillard reaction. *Food Chemistry* **177**: 248-257.

Pesic, M.B., Barac, M.B., Stanojevic, S.P. et al. (2014). Effect of pH on heat-induced casein-whey protein interactions: a comparison between caprine milk and bovine milk. *International Dairy Journal* **39** (1): 178-183.

Qi, X.L., Brownlow, S., Holt, C. et al. (1995). Thermal denaturation of β-lactoglobulin: effect of protein concentration at pH 6.75 and 8.05. *Biochimica et Biophysica Acta (BBA)-Protein Structure and Molecular Enzymology* **1248** (1): 43-49.

Qi, P.X., Xiao, Y., and Wickham, E.D. (2017). Stabilization of whey protein isolate (WPI) through interactions with sugar beet pectin (SBP) induced by controlled dry-heating. *Food Hydrocolloids* **67**: 1-13.

Rich, L.M. and Foegeding, E.A. (2000). Effects of sugars on whey protein isolate gelation. *Journal of Agricultural and Food Chemistry* **48** (10): 5046-5052.

Roefs, S.P.F.M. and Kruif, K.G.D. (1994). A model for the denaturation and aggregation of β-lactoglobulin. *European Journal of Biochemistry* **226** (3): 883-889.

Sabato, S.F., Nakamurakare, N., and Sobral, P. (2007). Mechanical and thermal properties of irradiated films based on Tilapia (Oreochromis niloticus) proteins. *Radiation Physics and Chemistry* **76** (11-12): 1862-1865.

Sanchez, C., Schmitt, C., Babak, V.G. et al. (2010). Rheology of whey protein isolate-xanthan mixed solutions and gels. Effect of pH and xanthan concentration. *Molecular Nutrition & Food Research* **41** (6): 336-343.

Sarbon, N.M., Badii, F., and Howell, N.K. (2015). The effect of chicken skin gelatin and whey protein interactions on rheological and thermal properties. *Food Hydrocolloids* **45**: 83-92.

Schmid, M., Sängerlaub, S., Wege, L. et al. (2015). Properties of transglutaminase crosslinked whey protein isolate coatings and cast films. *Packaging Technology and Science* **27** (10): 799-817.

Schokker, E.P., Singh, H., and Creamer, L.K. (2000). Heat-induced aggregation of β-lactoglobulin A and B with α-lactalbumin. *International Dairy Journal* **10** (12): 843-853.

Shen, X., Shao, S., and Guo, M. (2017a). Ultrasound-induced changes in physical and functional properties of whey proteins. *International Journal of Food Science and Technology* **52** (2): 381-388.

Shen, X., Zhao, C.H., and Guo, M.R. (2017b). Effects of high intensity ultrasound on acid-induced gelation properties of whey protein gel. *Ultrasonics Sonochemistry* **39**: 810-815.

Singh, H. and Havea, P. (2003). Thermal denaturation, aggregation and gelation of whey proteins. In: *Advanced Dairy Chemistry-1 Proteins*, 1261-1287. Berlin: Springer.

Solar, O. and Gunasekaran, S. (2010). Rheological properties of rennet casein-whey protein gels prepared at different mixing speeds. *Journal of Food Engineering* **99** (3): 338-343.

Spahn, G., Baeza, R., Santiago, L. et al. (2008). Whey protein concentrate/λ-carrageenan systems: effect of processing parameters on the dynamics of gelation and gel properties. *Food Hydrocolloids* **22** (8): 1504-1512.

Spotti, M.J., Perduca, M.J., Piagentini, A. et al. (2013). Gel mechanical properties of milk whey protein-dextran conjugates obtained by Maillard reaction. *Food Hydrocolloids* **31** (1): 26-32.

Spotti, M.J., Martinez, M.J., Pilosof, A.M. et al. (2014a). Influence of Maillard conjugation on structural characteristics and rheological properties of whey protein/dextran systems. *Food Hydrocolloids* **39**: 223-230.

Spotti, M.J., Martinez, M.J., Pilosof, A.M. et al. (2014b). Rheological properties of whey protein and dextran conjugates at different reaction times. *Food Hydrocolloids* **38**: 76-84.

Stender, E., Koutina, G., Almdal, K. et al. (2018). Isoenergic modification of whey protein structure by denaturation and crosslinking using transglutaminase. *Food & Function* **9** (2): 797-805.

Stone, A.K. and Nickerson, M.T. (2012). Formation and functionality of whey protein isolate-(kappa-, iota-, and lambda-type) carrageenan electrostatic complexes. *Food Hydrocolloids* **27** (2): 271-277.

Sun, W.W., Yu, S.J., Yang, X.Q. et al. (2012). Study on the rheological properties of heat-induced whey protein isolate-dextran conjugate gel. *Food Research International* **44** (10): 3259-3263.

Syrbe, A., Bauer, W.J., and Klostermeyer, H. (1998). Polymer ccience concepts in dairysystems—an overview of milk protein and food hydrocolloid interaction. *International Dairy Journal* **8** (3): 179-193.

Tammineedi, C.V., Choudhary, R., Perez-Alvarado, G.C. et al. (2013). Determining the effect of UV-C, high intensity ultrasound and nonthermal atmospheric plasma treatments on reducing the allergenicity of α-casein and whey proteins. *LWT-Food Science and Technology* **54** (1): 35-41.

Tobin, J.T., Fitzsimons, S.M., Kelly, A.L. et al. (2010). Microparticulation of mixtures of whey protein and inulin. *International Journal of Dairy Technology* **63** (1): 32-40.

Tolstoguzov, V. (1991). Functional properties of food proteins and role of protein-polysaccharide interaction. *Food Hydrocolloids* **4** (6): 429-468.

Tolstoguzov, V. (1995). Some physico-chemical aspects of protein processing in foods. Multicomponent gels. *Food Hydrocolloids* **9** (4): 317-332.

Turgeon, S.L. and Beaulieu, M. (2001). Improvement and modification of whey protein gel texture using polysaccharides. *Food Hydrocolloids* **15** (4-6): 583-591.

Tziboula, A. and Horne, D. (1999). Influence of whey protein denaturation on κ-carrageenan gelation. *Colloids and Surfaces B: Biointerfaces* **12** (3-6): 299-308.

Vachon, C., Yu, H.L., Yefsah, R. et al. (2000). Mechanical and structural properties of milk protein edible films cross-linked by heating and γ-irradiation. *Journal of Agricultural and Food Chemistry* **48** (8): 3202-3209.

Vardhanabhuti, B. and Foegeding, E.A. (1999). Rheological properties and characterization of polymerized whey protein isolates. *Journal of Agricultural and Food Chemistry* **47** (9): 3649-3655.

Vasbinder, A.J., Alting, A.C., and de Kruif, K.G. (2003). Quantification of heat-induced casein-whey protein interactions in milk and its relation to gelation kinetics. *Colloids and Surfaces B: Biointerfaces* **31** (1-4): 115-123.

Vasbinder, A.J., van de Velde, F., and de Kruif, C.G. (2004). Gelation of casein-whey protein mixtures. *Journal of Dairy Science* **87** (5): 1167-1176.

Verheul, M., Spfm, R., and Kgde, K. (1998). Kinetics of heat-induced aggregation of beta-lactoglobulin. *Journal of Agricultural and Food Chemistry* **46** (3): 896-903.

Visschers, R.W. and de Jongh, H.H. (2005). Disulphide bond formation in food protein aggregation and gelation. *Biotechnology Advances* **23** (1): 75-80.

Vu, K.D., Hollingsworth, R.G., Salmieri, S. et al. (2012). Development of bioactive coatings based on γ- irradiated proteins to preserve strawberries. *Radiation Physics and Chemistry* **81** (8): 1211-1214.

Walkenström, P. and Hermansson, A. (1994). Mixed gels of fine-stranded and particulate networks of gelatin and whey proteins. *Food Hydrocolloids* **8** (6): 589-607.

Walkenström, P. and Hermansson, A. (1996). Fine-stranded mixed gels of whey proteins and gelatin. *Food Hydrocolloids* **10** (1): 51-62.

Walkenström, P. and Hermansson, A. (1997a). High-pressure treated mixed gels of gelatin and whey proteins. *Food Hydrocolloids* **11** (2): 195-208.

Walkenström, P. and Hermansson, A. (1997b). Mixed gels of gelatin and whey proteins, formed by combining temperature and high pressure. *Food Hydrocolloids* **11** (4): 457-470.

Wang, W., Bao, Y., and Chen, Y. (2013). Characteristics and antioxidant activity of water-soluble Maillard reaction products from interactions in a whey protein isolate and sugars system. *Food Chemistry* **139** (1-4): 355-361.

Wang, X.B., Wang, C.N., Zhang, Y.C. et al. (2018). Effects of gamma radiation on microbial, physicochemical, and structural properties of whey protein model system. *Journal of Dairy Science* **101** (6): 4879-4890.

Weinbreck, F., Nieuwenhuijse, H., Robijn, G.W. et al. (2004). Complexation of whey proteins with carrageenan. *Journal of Agricultural and Food Chemistry* **52** (11): 3550-3555.

Wijaya, W., Van der Meeren, P., and Patel, A.R. (2017). Cold-set gelation of whey protein isolate and low-methoxyl pectin at low pH. *Food Hydrocolloids* **65**: 35-45.

Wooster, T.J. and Augustin, M.A. (2007). Rheology of whey protein-dextran conjugate films at the air/water interface. *Food Hydrocolloids* **21** (7): 1072-1080.

Xie, Y.R. and Hettiarachchy, N.S. (1998). Effect of xanthan gum on enhancing the foaming properties of soy protein isolate. *Journal of the American Oil Chemists Society* **75** (6): 729-732.

Yamul, D.K. and Lupano, C.E. (2005). Whey protein concentrate gels with honey and wheat flour. *Food Research International* **38** (5): 511-522.

Yazdi, S.R. and Corredig, M. (2012). Heating of milk alters the binding of curcumin to casein micelles. A fluorescence spectroscopy study. *Food Chemistry* **132** (3): 1143-1149.

Yong, Y.H. and Foegeding, E.A. (2008). Effects of caseins on thermal stability of bovine β-lactoglobulin. *Journal of Agricultural and Food Chemistry* **56** (21): 10352-10358.

Zhang, S. and Vardhanabhuti, B. (2014). Acid-induced gelation properties of heated whey proteinpectin soluble complex (Part II): effect of charge density of pectin. *Food Hydrocolloids* **39** (2): 95-103.

Zhang, S., Zhang, Z., Lin, M. et al. (2012). Raman spectroscopic characterization of structural changes in heated whey protein isolate upon soluble complex formation with pectin at near neutral pH. *Journal of Agricultural and Food Chemistry* **60** (48): 12029-12035.

Zhang, S., Hsieh, F.H., and Vardhanabhuti, B. (2014). Acid-induced gelation properties of heated whey protein-pectin soluble complex (Part I): effect of initial pH. *Food Hydrocolloids* **36**: 76-84.

Zhu, D., Damodaran, S., and Lucey, J.A. (2008). Formation of whey protein isolate (WPI)-dextran conjugates in aqueous solutions. *Journal of Agricultural and Food Chemistry* **56** (16): 7113-7118.

Zirbel, F. and Kinsella, J. (1988). Effects of thiol reagents and ethanol on strength of whey protein gels. *Food Hydrocolloids* **2** (6): 467-475.

第5章 乳清蛋白质的营养特性

Kelsey M. Mangano[1], Yihong Bao[2], and Changhui Zhao[3]

1. Department of Biomedical and Nutritional Science, University of Massachusetts, Lowell, MA, USA
2. Food Science and Engineering, College of Forestry, Northeast Forestry University, Harbin, People's Republic of China
3. Department of Food Science, College of Food Science and Engineering, Jilin University, Changchun, People's Republic of China

作为乳中的一类主要蛋白质，乳清蛋白质及其衍生物对人类健康和疾病具有潜在的医疗作用。例如，乳清蛋白质具有预防和治疗糖尿病（Gunnerud et al., 2012; Hoefle et al., 2015; Wildová and Anděl, 2013）、心血管疾病（Kawase, 2000）、肝病（Tanaka et al., 1999; Watanabe et al., 2000）、免疫相关疾病（Rusu et al., 2009; Shute, 2004）和肥胖症（Abreu et al., 2012; Panahi, 2014; Tina et al., 2010）的潜力。乳清蛋白质对健康的有益作用主要归因于其抗氧化特性。乳清蛋白质成分具有抗氧化活性。美拉德反应和提供半胱氨酸以合成谷胱甘肽（一种有效的抗氧化剂）是乳清蛋白质在食品中提供其潜在抗氧化性的两个重要指标。本章讨论了有关乳清蛋白质影响健康和疾病及其抗氧化活性的最新研究进展。

5.1 氨基酸组成：乳清蛋白质与母乳蛋白

乳清蛋白质是牛乳的两种主要蛋白成分之一，占牛乳蛋白的20%，其余的80%为酪蛋白。乳清蛋白质的所有组分都含有大量的必需氨基酸和支链氨基酸（BCAA）——亮氨酸、异亮氨酸和缬氨酸。乳清分离蛋白（WPI）中的亮氨酸含量（14 g/100 g）和总 BCAA 含量（26 g/100 g）高于乳蛋白（亮氨酸 10 g/100 g，BCAA 21 g/100 g）、蛋清蛋白（亮氨酸 8.5 g/100 g，BCAA 20 g/100 g）、大豆蛋白（亮氨酸 8 g/100 g，BCAA 18 g/100 g）和小麦蛋白（亮氨酸 7 g/100 g，BCAA 15 g/100 g）（Millward et al., 2008）。这些氨基酸是组织生长和修复的重要因子（Daenzer et al., 2001），亮氨酸是蛋白质代谢的关键氨基酸（Anthony et al., 2001）。乳清蛋白质中还含有丰富的含硫氨基酸，如半胱氨酸和甲硫氨酸，这些氨基酸可在细胞内转化为谷胱甘肽（GSH），具有增强免疫功能的效果（Grimble，2006）。与游离氨基酸相比，乳清蛋白质的氨基酸更容易被吸收和利用（Daenzer et al.，2001），这也表明从食物中摄取氨基酸比游离氨基酸对人体健康更有益。

与牛乳相比，人乳含有其独特的氨基酸分布。人乳中甲硫氨酸、苯丙氨酸和赖氨酸的含量比牛乳中的低，而半胱氨酸和色氨酸含量却高于牛乳。鉴于这些氨基酸的差异，将乳清蛋白质和乳蛋白混合（如60%的乳清蛋白质和40%的乳蛋白混合物）可实现氨基酸配比优化平衡。乳清蛋白质的添加将部分补偿乳蛋白中低丰度的色氨酸和半胱氨酸，但同时会使苏氨酸和赖氨酸的含量过高（Dupont，2003）。β-乳球蛋白（β-LG）是干酪乳清的主要蛋白成分，也是牛乳中潜在的主要变应原，人乳中没有β-LG。因此，婴儿首次食用牛乳清蛋白质必须监测过敏反应。人乳与牛乳相比是独特的，因为它富含乳铁蛋白（Lf），该蛋白已被证实对婴儿具有强大的免疫学作用（Gregory and Walker，2013）。Lf可以被添加到乳清蛋白质婴儿配方粉中以模拟母乳中Lf的抗微生物效果。因为在致敏性和对健康影响方面存在差异，有必要了解乳清蛋白质和母乳中主要氨基酸及蛋白成分的差异。

5.2 乳清蛋白质中的支链氨基酸

乳清蛋白质的氨基酸中含有丰富的BCAA（Mcdonough et al.，1974；Da Silva et al.，2015），其中亮氨酸和缬氨酸约占总氨基酸含量的26%。BCAA是一类优质氨基酸，在消化道中溶解度高，因此与其他需要预先消化才能在肠道中可溶的蛋白质相比，BCAA的吸收速度更快。BCAA（亮氨酸、异亮氨酸和缬氨酸）占肌肉蛋白质中必需氨基酸的35%，占哺乳动物所需必需氨基酸的40%（Harper et al.，1984；Shimomura et al.，2004）。与摄入酪蛋白或大豆相比，乳清蛋白质能更有效地增加血浆必需氨基酸和BCAA的浓度（Tang et al.，2009）。餐后体循环中必需氨基酸的增加可以促进蛋白质合成。乳清蛋白质比酪蛋白能更有效地刺激健康老年男性餐后肌肉蛋白的积聚，这归因于乳清更快的消化、吸收动力学以及更高的亮氨酸含量（Bart et al.，2011）。尽管含有乳清蛋白质补充剂的总蛋白质和能量含量低于大豆制品，但与大豆相比，将乳清蛋白质掺入即食型辅助食品可以使中度急性营养不良儿童获得更高的恢复率并促进其生长（Stobaugh et al.，2016）。

BCAA是促进必需氨基酸合成肌肉蛋白质所必需的（Katsanos et al.，2006；Mobley et al.，2016）。此外，BCAA可以提供能量并在运动中发挥关键作用。BCAA在肌肉组织中的降解迅速，与其他氨基酸相比，BCAA产生腺苷三磷酸（ATP）或能量的能力更高。BCAA促进ATP生成的机制可能是胰岛素分泌增加或相应代谢增强（Tatpati et al.，2007）。补充乳清蛋白质和亮氨酸所提高人体机能的效果超出了抵抗力训练和补充碳水化合物的效果（Coburn et al.，2006）。因此，BCAA 是运动后肌肉蛋白合成和恢复的重要营养因子。由此推断，乳清蛋白质是防止肌肉损失以及合成新肌肉蛋白的理想补充剂。

5.3 乳清蛋白质衍生物

乳清中的主要蛋白质有β-LG、α-LA、GMP、Ig、BSA、Lf、LP 和蛋白胨（PP）（Yadav et al.，2015）。乳清蛋白质对健康有很多益处，包括：①提高体能、锻炼后的恢复和预防肌肉退化（Farnfield et al.，2009a，b；Ha and Zemel，2003；Tipton et al.，2004）；②饱腹感和体重管理（Marsset-Baglieri et al.，2014；Zemel and Zhao，2009）；③心血管健康（Pins and Keenan，2006）；④预防癌症（Bounous，2000；Bounous et al.，1991）；⑤促进伤口护理和修复（Badr，2012，2013；Badr et al.，2012；Ebaid et al.，2011，2013）；⑥抗菌活性（Bounous et al.，1993）；⑦对婴儿营养的益处（Exl，2001）；⑧健康老龄化（Karelis et al.，2015；Paddon-Jones et al.，2015）。此外，乳清蛋白质是生物活性肽的重要来源，生物活性肽是指能提高人体健康的氨基酸序列。这些肽具有抗氧化、抗高血压和抗微生物等作用。生物活性肽可以通过乳清蛋白质的发酵或酶水解来制备。改善这些生物肽活性和浓度（de Castro and Sato，2015）的技术有蒸发、喷雾干燥或冷冻干燥等，可将乳清蛋白质转化为乳清粉、WPC、WPI、乳清渗透物、乳糖等产品。乳清蛋白质衍生物之所以受到重视，是因为它们除了具有改善食品质量的功能外，还具有潜在的促进健康的作用。乳清蛋白质的另一个重要应用是作为一种潜在的可生物降解的食品包装材料，主要是因为它们具有良好的机械强度以及优良的氧气、油脂和气味阻隔能力。但是，它们的防潮性能很差，而且材料很脆。增塑剂如甘油的掺入可以降低其脆性（Kokoszka et al.，2010）。随着研究和技术的发展，乳清蛋白质衍生物在人类健康、生物工程和生物技术方面具有广阔的应用前景。

5.4 乳清蛋白质的致敏性和消化率

测定蛋白质质量的方法有很多。常用的方法包括蛋白质效率比、生物学价值、净蛋白质利用率和蛋白质消化率校正氨基酸评分（PDCAAS）。PDCAAS 考虑了蛋白质的真实粪便消化率，是目前最为人们接受和广泛使用的测量营养中蛋白质价值的方法（Schaafsma，2000）。乳清蛋白质的 PDCAAS 得分为完美的 1.00 分，与鸡蛋、牛奶和酪蛋白的得分相似（U. S. Dairy Export Council，1999；Sarwar，1997）。得分为 1.00 表示乳清蛋白质易消化，可被人体有效利用。尽管乳清蛋白质在其消化率和利用率方面符合最高标准，但乳清蛋白质的组分（如β-LG）与过敏症有关，特别是在儿童中，2%~7%的儿童对牛乳不耐受（Høst，2002）。与乳清蛋白质相关的过敏症状包括异位性皮肤炎（Botteman and Detzel，2015）、呼吸窘迫（哮喘、鼻炎、喘息、咳嗽和喉头水肿）（Hochwallner et al.，2014）、胃肠道不适（腹泻、腹痛、恶心、呕吐）（Kattan et al.，2011）和敏感婴儿的过敏反应（Ameratunga and Woon，2010）。人们发现某些加工技术可以降低乳清蛋白质的过敏性。90℃以上的热处理可以降低WPI中α-LA和β-LG的抗原性（Bu et al.，2009）。酶促水解还降低了乳清蛋白质的致敏性（Zhang et al.，2012）。

5.5 乳清蛋白质组分的治疗特性

各乳清蛋白质组分对疾病有多种治疗作用。临床试验表明，乳清蛋白质在治疗癌症、人类免疫缺陷病毒（HIV）感染、乙型肝炎、心血管疾病、骨质疏松症等方面具有积极影响，并可作为抗菌剂使用（Marshall，2004）。上述这些功能主要来自于特定的乳清蛋白质组分。

5.5.1 糖尿病

2型糖尿病的患病率在全球范围内呈上升趋势，这与体育活动减少和饮食质量差密切相关（Sigal et al.，2006；Hansen et al.，2010）。高血糖指数与低胰岛素指数（II）高度相关，是2型糖尿病的危险因素。乳制品已多次被证明可以增强胰岛素分泌和降低血糖指数（GI）（Nilsson，2006）。有理论认为，这种特性是由牛乳中的蛋白质成分所致，因为乳糖和乳脂肪都不是胰岛素原。与酪蛋白、牛乳和许多其他蛋白质相比，乳清蛋白质具有更强的促胰岛素能力（Wildová and Anděl，2013）。与人乳和牛乳相比，重组牛乳清可以诱导胰岛素显著增加（Gunnerud et al.，2012）。牛乳清蛋白质被证明可刺激健康个体和糖尿病前期患者的胰岛素分泌并促进餐后血糖调节（Hoefle et al.，2015）。此外，还有人提出，乳清蛋白质通过其BCAA对胰腺β细胞的直接作用以及通过激活肠促胰岛素激素来增加胰岛素的分泌（Wildová and Anděl，2013）。乳清蛋白质具有的促胰岛素作用是由于摄入后血浆中几种必需氨基酸增加。这些氨基酸包括亮氨酸、异亮氨酸、缬氨酸、赖氨酸和苏氨酸（Mikael et al.，2007），其中亮氨酸是一种已知的强促胰岛素分泌剂。另一个机制是乳清蛋白质能够提高血清中胰岛素营养多肽（GIP）和胰高血糖素样肽1（GLP-1）的水平。GLP-1和GIP都是强促胰岛素分泌激素。此外，乳清蛋白质中含量丰富的膳食BCAA也能促进GLP-1从肠细胞中释放。因此，乳清蛋白质可以通过促进胰岛素分泌而为糖尿病患者提供饮食干预。

肥胖和高血糖是诱发糖尿病的重要危险因素。乳蛋白在控制肥胖方面引起了广泛关注，因为高乳制品摄入量和低体重之间有着密切的联系（Abreu et al.，2012）。推断乳蛋白通过胰岛素依赖和非依赖性的方式改善血糖控制来降低短期食欲和食物摄入（Panahi，2014；Tina et al.，2010）。乳清蛋白质可通过三种机制降低血糖，即快速消化、高含量的 BCAA 和快速胰岛素刺激。完整的乳清蛋白质、水解乳清蛋白质和 BCAA 对血浆胰岛素和血糖水平的影响相似。然而，BCAA 混合物不能再现乳清蛋白质对肠肽的影响（肠肽涉及控制葡萄糖和胃排空作用），这可能与胰岛素依赖机制有关（Nilsson et al.，2007）。因此，乳清蛋白质在糖尿病的治疗中具有极大潜力。

5.5.2 癌症

乳清蛋白质对癌症的预防和治疗作用在动物模型中得到了广泛的研究（Smithers et al.，1998）。以乳清蛋白质为唯一蛋白源的饮食在减少大鼠化学性乳腺肿瘤的发生率和多样性方面比富含大豆的饮食更有效（Hakkak et al.，2000）。与酪蛋白相比，乳清蛋白质对大鼠在患结肠癌方面具有更好的保护作用（Hakkak et al.，2001）。几种乳清源蛋白质已被证实有助于这种抗肿瘤作用。在最近的一项研究中，牛 Lf 给药一个月或半年可显著抑制氧化偶氮甲烷引发的大鼠结肠癌的发生和发展（Sekine et al.，1997；Tsuda et al.，1998）。牛 Lf 在小鼠模型中也显示出对转移的抑制作用（Yoo et al.，1997）。此外，乳清蛋白质成分 BSA 也能抑制人乳腺癌细胞的生长（Laursen et al.，1989）。

乳清蛋白质防癌作用的一个重要机制是其抗氧化性。乳清蛋白质水解物是一种由多肽和氨基酸组成的强抗氧化剂。乳清蛋白质的抗氧化活性来自于几种氨基酸残基，典型的有酪氨酸、甲硫氨酸、组氨酸、赖氨酸和色氨酸，因为这些氨基酸能够螯合金属。乳清蛋白质水解后得到的一些二肽或三肽在清除氧化剂方面发挥关键作用（Peña-Ramos and Xiong，2001）。与酪蛋白相比，乳清蛋白质的半胱氨酸含量更高。乳清蛋白质的抗氧化性可能取决于半胱氨酸和/或谷氨酰半胱氨酸基团的含量。充足的半胱氨酸供应是维持细胞内抗氧化剂谷胱甘肽（GSH）的必要条件。富含半胱氨酸的 WPI 可以通过提供细胞内谷胱甘肽合成所需的前体物质来提高谷胱甘肽水平（Micke et al.，2001）。

谷胱甘肽的合成被认为是乳清蛋白质预防癌症的重要机制之一。由于半胱氨酸是谷胱甘肽合成的关键限制氨基酸（图 5.1），所以乳清蛋白质是补充谷胱甘肽的有效而安全的半胱氨酸供体。因此，乳清蛋白质可以为谷胱甘肽体内平衡提供足够的蛋白质营养。众所周知，谷胱甘肽在抗氧化防御、营养物质代谢和调节细胞事件如基因表达、DNA 和蛋白质合成、细胞增殖和凋亡、信号转导、细胞因子的产生和免疫反应以及蛋白质谷胱甘肽修饰等方面发挥重要作用。谷胱甘肽缺乏引起的氧化应激可以导致衰老以及增加包括癌症在内的许多疾病的风险（Wu et al.，2004）。乳清蛋白质能够增加正常组织细胞中的谷胱甘肽含量，但会减少肿瘤细胞中谷胱甘肽的含量，从而使肿瘤细胞更容易受到化疗的影响（Kennedy et al.，1995）。

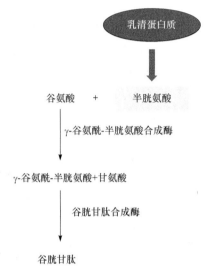

图 5.1　乳清蛋白质在谷胱甘肽合成中的作用

乳清蛋白质抗肿瘤作用的另一种解释是铁的调节作用。铁被认为具有致癌性，因为它能促进羟基自由基的形成，抑制宿主防御并促进癌细胞增殖（Weinberg，1996）。铁保留蛋白有助于癌症的预防和治疗。Lf 被认为是铁结合蛋白之一。与此相一致的是，强化乳清蛋白质的营养保健品可有效提高各类癌症晚期患者自然杀伤细胞的功能和其他免疫参数（如血红蛋白）（See et al.，2002）。

5.5.3　肝病

乳清蛋白质可以保护肝脏免受损伤和感染。丙型肝炎病毒是一种能引起丙型肝炎并增加肝硬化、某些癌症和淋巴瘤风险的病毒。乳清蛋白质成分对丙型肝炎患者有积极影响。牛 Lf 是铁转运蛋白家族的一员，可以有效地防止丙型肝炎病毒在培养人肝细胞中的感染（Ikeda et al.，1998）。慢性肝炎患者补充八周的 Lf 可以降低血清丙氨酸转氨酶和丙型肝炎病毒 RNA 的浓度（Tanaka et al.，1999）。但其与常规治疗相结合的最佳剂量还需要进一步的研究来确定。乳清蛋白质可以改善慢性乙型肝炎患者的肝功能，进一步表明了其在临床上作为辅助疗法的潜力（Watanabe et al.，2000）。

谷胱甘肽缺乏会导致肝脏损伤，甚至引起脂肪性肝炎。乳清蛋白质由于其富含半胱氨酸可以减轻由 GSH 缺乏产生的肝损伤（Balbis et al., 2009）。长期以来氧化应激被认为是造成肝损伤和疾病恶化的主要原因，乳清蛋白质可以改善氧化应激（Bonnefont-Rousselot et al., 2006）。事实上，口服 WPI 可以改善非酒精性脂肪性肝炎患者的肝脏生化指标，增加血浆 GSH，提高抗氧化能力并减少肝性脂肪变性（Chitapanarux et al., 2009）。

5.5.4 心血管疾病

心血管疾病是各类心脏和/或血管疾病的统称。食源性保健食品在心血管疾病的药物治疗中具有广阔的应用前景。乳清蛋白质可能会减少心血管疾病的危险因素，如添加了 WPC 的发酵乳改善了人的血脂水平，降低了其收缩压，这表明乳清蛋白质有可能降低患心血管疾病的风险（Kawase, 2000）。主要的心血管疾病与肥胖密切相关（Lavie et al., 2009）。超重的绝经后妇女摄入乳清蛋白质数小时后，与摄入酪蛋白相比，血浆甘油三酯降低了约 30%（Pal et al., 2010）。超重和肥胖的男性持续几周在主餐前 30 min 摄入乳清蛋白质，与大豆蛋白相比，乳清蛋白质对血压、快速血糖水平和血脂的影响更为有益（Tahavorgar et al., 2015; Golzar et al., 2012）。

高血压患者通常通过抑制血管紧张素转换酶（ACE）进行药物治疗。ACE 是调节血压的关键酶，是影响心血管疾病的重要因素。抑制 ACE 活性可以阻止血管紧张素 I 向血管紧张素 II 的转化。因此，乳清蛋白质降低心血管疾病风险的机制可能归因于源于乳清蛋白质的具有 ACE 抑制作用的肽（Pan et al., 2012; Cvander et al., 2002; Zhao et al., 2010）。该机制所涉及的分子途径还需要进一步的研究来探索。

5.5.5 免疫系统疾病

免疫球蛋白是一种重要的蛋白质，通常具有抗体活性。主要的免疫球蛋白 IgG 约占成人抗体的 75%。IgG 在子宫内通过脐带血和母乳喂养从母亲转移到孩子体内，成为孩子的第一道免疫防线。另一种免疫球蛋白 IgA 从母乳转移至新生儿的消化道。因此，母乳喂养的一个优点是比奶粉喂养的孩子提供更多的免疫因子，从而更好地防止革兰氏阴性肠杆菌的肠道定植（Bonang et al., 2000）。

乳清含有大量的免疫球蛋白，约占总蛋白质的 10%~15%。牛乳中免疫球蛋白可能有调节人体免疫反应的潜力。牛乳中 IgG 含量在 300 μg/mL 可抑制 IgG、IgA 和 IgM 的合成（抑制率为 96%~98%），也可抑制人细胞的有丝分裂反应。中度加热（63℃，30 min）可增强其抑制作用（Kulczycki and Macdermott, 1985）。因此，牛乳很可能赋予人类免疫力。原料乳含有特异性抗体，这些抗体对细菌性病原体的脂多糖具有足够的免疫活性（Losso et al., 1993）。生鲜牛乳和巴氏杀菌牛乳中的 IgG 抗体对轮状病毒感染小鼠具有保护作用（Yolken et al.，1985）。

Lf 在人体免疫系统中也起着至关重要的作用。例如，在广泛使用的三联疗法（雷贝拉唑、克拉霉素、替硝唑）的基础上，应用 Lf 治疗 7 天可有效根除幽门螺杆菌（*H. pylori*）（Di Mario et al.，2003）。另一个很好的例子是使用 Lf 治疗被甲类链球菌感染的患者，其临床与其他兼性胞内细菌病原体具有相似的行为，这些病原体进入宿主细胞并在宿主细胞中生存，且其未在细胞内复制，不会引起宿主防御，当周围条件允许时导致感染复发。牛乳 Lf 已被证明能够在体外和体内减少甲类链球菌的侵袭（Ajello et al.，2002）。

补充乳清蛋白质两周可大大增加外周血单个核细胞的淋巴细胞增殖，这有助于在剧烈训练或比赛期间维持运动员的健康（Shute，2004）。乳清蛋白质通过 β-LG 和 α-LA 可以增强或"启动"人类中性粒细胞对后续刺激的反应（Rusu et al.，2009）。综上所述，乳清蛋白质在调节免疫系统方面或许有很大的潜力，这对许多免疫相关疾病的治疗将会非常有帮助。

5.6 乳清蛋白质的抗氧化特性

乳清蛋白质除了具有很好的功能特性（如溶解性、起泡性、乳化性和凝胶化）外，还具有一些很好的生理活性，如抗氧化活性、抗菌活性、免疫调节作用、抗肿瘤活性和促肌肉生长（Yadav et al.，2015）。乳清蛋白质具有抗衰老、增强机体免疫力的作用，是目前抗氧化活性最强的蛋白质。简而言之，人体内的生化反应会产生活性氧（ROS）和自由基，加剧 DNA、蛋白质和细胞小分子等生物分子的氧化损伤，如果不及时治疗，会导致多种疾病。抗氧化剂可以阻止自由基的有害作用，清除自由基并消除其对细胞成分的破坏作用。来自乳清蛋白质的天然抗氧化剂，包括多肽及其改性产品，可以提高血浆的抗氧化能力，并降低某些疾病的风险，如癌症、心脏病和卒中（Yao et al.，2016）。乳清蛋白质产品还被证明可以调节脂肪并增强免疫功能（Schröder et al.，2017）。因此，乳清蛋白质的抗氧化活性可用于疾病预防，乳清蛋白质及其衍生物的抗氧化活性在随后的小节中描述。

5.6.1 乳清蛋白质的抗氧化活性

虽然乳清蛋白质含有许多蛋白质组分，但其总的抗氧化活性已被广泛研究。总乳清蛋白质不需要工业分离，可以作为一个整体蛋白质抗氧化剂应用于食品工业来降低成本。本节将讨论两种方法：美拉德反应和提供半胱氨酸蛋白合成 GSH。

美拉德反应能增加总乳清蛋白质的抗氧化能力。美拉德反应不仅影响食品的颜色、气味、味道和质量，而且影响乳清蛋白质在食品加工过程中的营养特性。

美拉德反应产生一系列复杂的化合物，如黑色素、还原酮和一系列具有抗氧化活性的氮、硫挥发性杂环化合物。温度是影响美拉德反应的重要因素，随着反应温度和反应时间的增加，美拉德反应的速率大大加快，其产物的抗氧化活性也随之发生变化。在 Wang 等（2013）的一项研究中，从 WPI 中提取的美拉德反应产物（MRPs）随着 pH 的增加呈现出最高的褐变程度、还原力和 2, 2-二苯基-1-苦基肼（DPPH）自由基清除活性。因此，从 WPI 和木糖反应体系获得的 MRPs 比未处理的乳清蛋白质具有更高的抗氧化活性（Wang et al.，2013）。此外，Zhang 等还证明，乳清蛋白质-木糖经干热（相对湿度 60℃和 79%）0～7 天后，其颜色（红色和黄色）、紫外可见吸收和荧光强度均随反应时间的延长而增加。与葡萄糖结合后，WPI 的热稳定性显著提高，随着反应时间的延长，WPI 的还原能力和二铵盐（ABTS）自由基清除活性显著增加（Liu et al.，2014）。Chawla 和他的同事们测试了γ射线照射对乳清蛋白质粉 MRPs 的影响，MRPs 的还原力和螯合铁能力随辐照剂量的增加而增强，游离氨基和乳糖的减少表明辐照处理导致糖基化蛋白质形成（Chawla et al.，2009）。这些研究表明，在特定的加热和 pH 处理下可提高总乳清蛋白质的抗氧化活性。

乳清蛋白质的抗氧化能力也可以通过添加富含半胱氨酸的蛋白质来提高，半胱氨酸调节谷胱甘肽的含量，饮食中增加富含半胱氨酸的乳清蛋白质可促进谷胱甘肽的生物合成。与酪蛋白对照组相比，服用乳清粉补充剂 30 天的受试者的肌肉功能（通过腿部等速运动循环评估）有所改善。富含半胱氨酸的 WPI 在健康和患病的哺乳动物大脑中也有抗氧化作用。WPI 补充剂（免疫性）可能是一种治疗精神分裂症的安全和有效的方式（Song et al.，2017）。因为氧化应激会导致肌肉疲劳，所以增加细胞内 GSH 的生物合成及其抗氧化活性能提高运动成绩。此外，艾滋病患者食用乳清补充剂可有效提高血浆谷胱甘肽浓度，虽然这种变化的最终效果尚不清楚（Ha and Zemel，2003）。乳清蛋白质有益人体健康，可用于治疗氧化应激相关疾病（El-Desouky et al.，2017）。

5.6.2 乳清蛋白质组分的抗氧化活性

1. β-乳球蛋白的抗氧化活性

β-LG 约占乳清蛋白质的 50%。最新研究表明，β-LG 可以结合维生素 A、维生素 E 等脂溶性相对较小的分子，β-LG 还具有一定的生物活性，如抑制 ACE、抗癌、抗微生物活性、降胆固醇、调节代谢等生理作用。β-LG 的改性包括与硒、姜黄素、多酚化合物的结合及美拉德反应（Wu et al.，2018）。

硒缺乏可引起多种疾病，如大骨节病、慢性退行性疾病和骨骼肌肌病。这些缺陷在某些情况下还可能导致癌症和免疫功能障碍。在真空和低温条件下用β-LG和二氧化硒合成的新型有机硒化合物具有抗癌活性（Zheng et al.，2016b）。硒偶联β-LG 可以合成具有抗肿瘤活性的 Se-β-LG 复合物（Zheng et al.，2016a）。β-LG 纳米粒子还用作营养载体以增强黏液黏附、跨膜渗透和细胞摄取。制备的β-LG-果胶纳米粒可将抗癌铂络合物转移到结肠癌组织中，是一种新颖而有效的口服给药载体。综上，Se-β-LG 复合物作为一种抗氧化剂可以有效缓解由于氧化还原失衡造成的细胞衰老，具有成为抗癌药物的潜力。

采用荧光和圆二色性（CD）光谱对β-LG 与茶、咖啡和可可的多酚提取物在胃肠道（GIT）pH 下的非共价相互作用进行研究（Sah et al.，2016）。采用体外胃蛋白酶和胰蛋白酶消化率以及复合物的 ABTS 自由基清除活性方法研究了多酚与β-LG 非共价结合的生物学意义。多酚-β-LG 体系在 GIT 的 pH 下稳定。对于富含酚酸的多酚提取物，pH 对亲和力的影响最大。强的非共价相互作用延缓了胃蛋白酶和胰蛋白酶对β-LG 的消化，并在中性 pH 时诱导β片层向α螺旋转变。所有多酚类物质在极酸性 pH 1.2 时都保护蛋白质二级结构。蛋白质-多酚的相互作用强度与蛋白质在胃部酸性条件下的降解半衰期（$R^2 = 0.85$）、蛋白质-多酚复合物的总抗氧化能力（TAOC）（$R^2 = 0.95$）呈正相关（Stojadinovic et al.，2013）。多酚类化合物与蛋白质络合物通过影响多酚类化合物的供电子能力和降低溶液中羟基的数量来影响多酚化合物的抗氧化活性。然而，由于络合物中多酚的活性周期延长，这种相互作用可能对多酚的总体抗氧化活性有益。

姜黄素是一种来源于姜黄属植物的酚类化合物，广泛分布于亚洲和中美洲的热带地区。有学者研究了姜黄素与牛β-LG 的结合，用傅里叶变换红外光谱和荧光光谱检测新形成复合物的抗氧化活性的改善（Li et al.，2013）。通过测定 ABTS、羟基自由基清除能力和总还原能力确定β-LG 对姜黄素抗氧化活性的影响，结果表明，当姜黄素与β-LG 结合时，可引起蛋白质结构的部分改变。姜黄素在 pH 6.0 和 7.0 时通过疏水作用结合到两个不同的蛋白质位点。一个蛋白质分子与一个姜黄素分子结合形成姜黄素-β-LG 复合物，两个结合位点到蛋白质中 Trp 残基的平均距离相似。本研究说明，姜黄素可能通过与β-LG 结合而提高其抗氧化活性。利用 Caco-2 细胞系进行的体外研究显示，与β-LG 稳定结合的姜黄素的生物利用度显著提高（Aditya et al.，2015）。共轭姜黄素-β-LG 复合物对健康的潜在益处还需要进一步研究。

超声处理是乳品工业中的一项新的处理技术，它在形成具有强抗氧化活性的 MRPs 方面显示出潜力。中性条件下，高强度超声波促进了β-LG 溶液的美拉德反应，超声分别处理β-LG 与葡萄糖、半乳糖、乳糖、果糖、核糖和阿拉伯糖混合物，发现有 MRPs 形成。圆二色谱分析显示其二级和三级结构有轻微变化。超声波法制备的 MRPs 具有 DPPH 清除活性以及较高的铁螯合活性和还原力。在非变性条件下，高强度超声能有效地促进水溶液中β-LG 美拉德反应产物的形成。β-LG 和核糖混合物在热处理过程中产生 MRPs，且 MRPs 的 pH 降低。在β-LG-核糖模拟体系中产生 MRPs 的氨基含量在第一个小时内明显下降，此后保持不变。在美拉德反应中，天然和非天然的β-LG 浓度降低并形成聚集体。MRPs 电泳结果显示，β-LG 单体通过分子间二硫键聚合，也通过其他共价键聚合。β-LG-核糖的 MRPs 具有较高的抗氧化活性，可作为天然抗氧化剂使用（Jiang and Brodkorb, 2012）。

人类免疫缺陷病毒 1 型（HIV-1）的异性传播是全球艾滋病持续流行的主要原因，应用化学屏障方法有望更好地控制这一流行病。一些研究报道，用 3-羟基邻苯二甲酸酐化学改性β-LG 形成的 3-羟基苯甲酰基-β-LG，在体外能有效抑制 HIV-1、HIV-2、猴免疫缺陷病毒、单纯疱疹病毒 1 型和 2 型以及沙眼衣原体感染（Wang et al., 2000）。经过改性的β-LG 有可能作为 HIV-1 感染的一种有效抑制剂。此外，β-LG 对轮状病毒复制的抑制作用与剂量有关（Chatterton et al., 2006）。但β-LG 改性的抗氧化机制尚不清楚，如上面提到的β-LG 与还原糖发生的美拉德反应，β-LG 与硒、姜黄素、多酚等的结合等，这些过程中产生的结构变化和抗氧化活性的改变机制还不清楚。因此，如何控制β-LG 改性反应过程是继续这一重要研究领域的关键。

2. α-乳白蛋白的抗氧化活性

α-LA 约占乳清蛋白质的 20%～25%，是人体必需氨基酸和支链氨基酸的优质来源，是唯一与钙结合的乳清蛋白质。它作为乳糖合成酶的调节亚基，在乳腺分泌细胞中具有重要的生物学功能，可以增加半乳糖基转移酶对葡萄糖的亲和力与特异性，从而促进乳糖合成（Al-Hanish et al., 2016）。在食品加工过程中，美拉德反应一般发生在高温条件下的溶液体系中。然而，目前只有少量已发表的数据可用（Yi et al., 2016）。

最近的一项研究评价了由α-LA 和α-LA-葡聚糖接枝物制备的姜黄素纳米复合物的特性，以及包封物对姜黄素的理化稳定性和抗氧化活性的影响（Jiménez-Castaño et al., 2007）。采用纳米包埋技术提高姜黄素的水分散性，包封率和载药量不受糖基化的影响。与单独α-LA 相比，α-LA-葡聚糖接枝物在环境应力作用下的稳定性显著提高。与游离姜黄素相比，被α-LA 和α-LA-葡聚糖接枝物包埋的姜黄素有更高的氧化稳定性。α-LA-葡聚糖纳米载体可作为姜黄素的优良递送体系在食品中广泛应用（Yi et al., 2016）。

研究α-LA 与阿洛糖（D-allose）以及两种食物糖（D-果糖，Fru；D-葡萄糖，Glc）的美拉德反应程度及其对α-LA 抗氧化性的影响，美拉德反应条件为温度 50℃、相对湿度 55%、反应时间 48 h。结果表明，D-allose 与α-LA 的美拉德反应速率和荧光显影速度均快于 Glc 和 Fru 修饰的α-LA。不同糖-蛋白结合物的荧光时间模式与生化活性之间具有良好的相关性（Sun et al., 2006）。这项研究表明，α-LA 与 D-allose 的聚合物在食品中可能具有重要的抗氧化活性。

通过 MRPs 的表征有助于控制末期 MRPs（对糖尿病和与年龄有关的心血管

疾病有害）的形成。但是，通过该控制过程也可以修饰蛋白质的功能。此外，用 D-allose 糖基化的蛋白质具有很强的抗氧化活性，可清除自由基并延缓由氧化引起的降解，可作为功能成分用于配方食品。

3. 乳铁蛋白的抗氧化活性

乳铁蛋白（Lf）具有结合和转移游离铁及其他二价金属离子的能力，这些离子可催化过氧化物自由基的形成。因此，Lf 可以有效地抑制金属离子催化的氧化过程，是一种理想的促进双歧杆菌增殖的抗氧化剂。食品成分的物理化学和结构特性是影响人体健康的基础，而不仅仅是其营养价值（Khan et al., 2013）。天然 Lf 具有抗氧化活性，而改性 Lf 也具有抗氧化活性。Lf 是转铁蛋白家族中的一员，每分子可与 2 个 Fe^{3+} 形成一个粉红色复合物。牛乳 Lf 具有广泛的药理作用，包括抗菌、抗氧化、免疫调节、抗转移和抗癌活性（Berkhout et al., 2002）。因为与蛋白质结合的铁不能作为催化剂产生羟基自由基，因此牛乳 Lf 被认为是一种抗氧化剂。牛乳 Lf 通过螯合铁离子、减少 H_2O_2 到 ·OH 的转化来抑制脂质过氧化反应。在早产儿的饮食中补充 Lf 也已证明可以减少铁诱导产生的氧化产物（Joubran et al., 2013）。

食品蛋白质和肽类可以作为抗氧化剂干扰自由基反应（Sila and Bougatef, 2016; Lykholat et al., 2016）。研究表明，将 Lf 用作乳化剂时，Lf（游离金属含量很少）也可以延迟乳液氧化（Liu et al., 2016a）。饲料中添加牛乳 Lf 对仔猪生产性能和抗氧化能力有很大影响，添加牛乳 Lf 能提高仔猪的日增重、日平均采食量和降低仔猪腹泻率，最高增重和最佳饲料转化率为 2500 mg/kg 牛乳 Lf，因此 Lf 有效地改善了仔猪的生长性能并增强了其抗氧化酶的活性和 mRNA 水平（Wang et al., 2008）。

4. 糖巨肽（GMP）的抗氧化活性

生物活性肽因其潜在的有益功效而受到广泛关注，这些有益功效包括抗菌、抗癌、抗高血压和免疫调节活性等。不同的研究也表明，肽可以干扰自由基反应，并作为抗氧化剂。乳清蛋白质还含有 GMP，这是一种 64 个氨基酸组成的可溶性肽，是由粗制凝乳酶（糜蛋白酶）水解κ-酪蛋白形成的。GMP 是酪蛋白巨肽（CMP）的糖基化形式，包含多种寡糖，主要有唾液酸（N-乙酰神经氨酸）、半乳糖胺和半乳糖（Chungchunlam et al., 2014）。肠道抗炎药 GMP 对大鼠脾细胞具有免疫调节作用（Requena et al., 2009），其作用机制可能与对淋巴细胞的作用有关。有学者研究了 GMP 在体外和体内对大鼠脾细胞的作用，分别用 BSA 和 Lf 进行比较。结果显示，GMP 能增加刀豆素 A，但不促进基底脾细胞的增殖。GMP 对 IFN-γ 分泌产生明显的抑制作用（70%），对 TNF-α 抑制作用较小（50%）。GMP 可能会限制或部分作用于肠道炎症的淋巴细胞（Requena et al., 2010），抑制了刀豆素 A 刺激的脾细胞中 TNF-α 和 IFN-γ 的表达，同时增强了静息细胞中 Foxp3 和 IL-10

的表达。

5. 牛血清白蛋白（BSA）的抗氧化活性

BSA 是一种具有多个药物结合位点的蛋白质，其价格低廉，具有非免疫原性和可生物降解性，已被用作基于纳米粒子的药物传递的基质（Fonseca et al., 2017）。BSA 通过结合天然活性成分来发挥其抗氧化活性。BSA 与木犀草素-氧化钒（Ⅳ）配合物在癌症治疗中的抗癌、抗氧化特性是目前的研究热点。研究显示，该配合物能提高木犀草素对羟基自由基的清除能力。对 MDAMB231 乳腺癌细胞株和 A549 肺癌细胞株抗肿瘤作用的评估结果显示，木犀草素通过静电力与 BSA 色氨酸基团相互作用（Naso et al., 2016）。木犀草素-BSA 复合物的形成过程是自发的，静电力是主导力，其配合物与 BSA 的相互作用是通过氢键和范德瓦耳斯力实现的。由于木犀草素及其配合物有细胞毒性，因此这种复合物作为一种潜在的药物，还需要通过体内试验研究。

BSA 与人血清白蛋白具有很高的序列同源性（76%），目前已经有很多关于人血清白蛋白和 BSA 糖基化的研究（Bodiga et al., 2013；Fan et al., 2018；Sadowska-Bartosz et al., 2015）。抗坏血酸增强了糖（尤其是葡萄糖）和 BSA 糖基化反应（Sadowska-Bartosz et al., 2015）。抗坏血酸既是促糖基化剂又是抗糖基化剂。在体外，抗坏血酸主要表现为促糖基化作用（Sadowska-Bartosz et al., 2015）。BSA 糖基化过程中伴随着氧化修饰。有研究合成了一类新型的 N 取代的糖胺衍生物，并对其抗惊厥、抗氧化活性以及与 BSA 的相互作用进行了评价。该研究通过最大电击癫痫发作（MESs）试验检查了合成化合物 4a~j 的抗惊厥活性，并通过小鼠转子试验确定了它们的神经毒性作用，还研究了这些化合物的结构活性关系。研究发现，化合物 4d、4g、4i 和 4j 对癫痫发作具有良好的保护作用（Prashanth et al., 2013）。然而，目前对 BSA 及其修饰产物抗氧化活性的研究还不全面。在将其作为食品抗氧化剂使用之前，还需要对其抗氧化剂活性进行进一步研究。

5.6.3 乳清蛋白质抗氧化肽

生物活性肽是一类具有特定生理功能的乳清蛋白质片段，是由蛋白质体外消化产生的，由数个到数十个氨基酸组成。目前，抗氧化的乳清蛋白质水解肽已被广泛研究。这些肽具有多种功能，包括清除自由基、提供氢、提供电子、螯合金属离子、猝灭单线态氧、分解过氧化物以及抑制脂质氧合等。

1. 水解肽的抗氧化活性

水解包括酸水解和酶水解。酶水解反应过程温和，水解产物安全，不会改变肽的营养价值。在适当的酶解条件下，乳清蛋白质可以产生抗氧化肽（Rocha et al., 2017）。水解肽制备工艺包括脱苦、脱色、分离和干燥等过程（Vavrusova et al., 2015）。

蛋白质水解产物是肽和氨基酸的复杂混合物，这些肽和氨基酸可能有不同的抗氧化性。因此，蛋白质及其水解产物的抗氧化性需要用不同的方法评估。有学者研究了WPI被不同蛋白酶（即胰蛋白酶、胃蛋白酶、碱性蛋白酶、复合蛋白酶、风味蛋白酶等）水解后的产物，其中碱性蛋白酶产生的水解产物显示出最高的抗氧化活性，并分离出七个具有强抗氧化活性的肽。抗氧化肽WYSL具有最强的DPPH自由基清除活性和超氧化物自由基清除活性，IC_{50}值分别为273.63 μmol/L和558.42 μmol/L（Brandelli et al.，2015）。

许多生物活性肽具有多种功能特性。例如，一些抗氧化肽也显示出ACE抑制活性。β-LG片段LQKW f（58～61）、LDDTYKK f（95～101）和FNPTQ f（151～155）包含Tyr（Y）和Trp（W）氨基酸，而这两种氨基酸体现出抗氧化活性，WPC被嗜热菌蛋白酶水解后呈现的抗氧化性能也主要归因于此（O'Loughlin et al.，2014）。肽LQKW对自发性高血压大鼠有ACE抑制活性和降压作用（Hernándezledesma et al.，2007）。Gu等的工作表明，嗜热菌蛋白酶消化牛Lf可以释放出有效的ACE抑制肽（Gu and Wu，2016）。综上，乳清蛋白质的水解条件对其抗氧化活性起着重要的作用。使用木瓜蛋白酶及其微生物源替代物（类木瓜蛋白酶活性）水解乳清蛋白质，评估这些条件对理化特性和生物活性的影响，结果表明，用相同酶产生的乳清蛋白质水解物（WPH）表现出相似的水解度，但它们的反相液相色谱的肽谱图不同。与不调节pH情况下产生的WPH相比，在恒定pH 7.0下木瓜蛋白酶水解乳清蛋白质的氧自由基吸收能力（ORAC）值明显更高，而pH调节对二肽基肽酶4（DPP-Ⅳ）的性质没有显著影响。与不调节pH情况下产生的WPH相比，在pH 7.0下类木瓜蛋白酶水解的乳清蛋白质显示出更高的ORAC活性和DPP-Ⅳ抑制特性。这项研究表明，WPH生成过程中的pH条件可能会影响肽的释放，致使WPH呈现出生物活性特性（Rani and Mythili，2014）。

Hongsprabhas等（2011）最近的一项研究调查了重构的WPH在热加工食品中作为抗褐变剂以及在生物系统中作为化学预防成分的潜在用途。在这项研究中，使用胰蛋白酶（EC 3.4.21.4）消化WPC或加热（80℃、30 min）乳清浓缩蛋白（HWPC）制备水解物。WPC和HWPC经胰蛋白酶水解后抗氧化能力从0.2 μmol提升至0.5 μmol[氧自由基吸收能力-荧光素（ORACFL）法]。HWPC中的MRPs对正常人肠道FHs 74 Int细胞和人上皮结直肠癌Caco-2细胞均具有细胞毒性。HWPC的IC_{50}约为3.18～3.38 mg/mL蛋白。当使用分别添加了WPH和HWPC的培养基培养这两种细胞时，它们均能够存活，而在含HWPC的培养基中以3.5 mg/mL浓度添加MRPs之后，细胞死亡。总的来说，这项研究表明了WPH和HWPH具有预防MRPs细胞毒性的能力。也有人指出，WPH和HWPH的ORACFL抗氧化能力必须足够高才能对MRPs的细胞毒性提供有效的化学预防作用。

从富含β-LG的WPC中获得的水解物已被证实可刺激人肠杯状细胞

HT29-MTX 中黏蛋白的分泌和黏蛋白 5AC 基因的表达（Martínez-Maqueda et al., 2013）。基于质谱的肽组学分析可以用于鉴定水解物中包含的肽。对于水解物活性的作用机制，尽管可能存在其他假说，但目前被广泛认为黏蛋白分泌反应的诱导是由 HT29-MTX 细胞中的 μ-阿片受体介导的。具有诱导黏蛋白分泌能力的蛋白水解物有望提高胃肠道的保护作用。

尽管目前已经对乳清源肽的抗氧化性进行广泛研究，但是对肽的结构-活性关系以及氨基酸和其他抗氧化剂化合物之间的协同和拮抗作用还需要进一步的研究。也需要开展进一步的工作来了解 WPI 在实际食品体系中的水解产物的抗氧化能力及其对食品感官特性的影响。从这个角度来看，乳清源产品中肽的抗氧化活性可以满足对旨在人体健康和食物品质方面的更天然抗氧化剂的不断增长的需求。

2. 改性乳清蛋白质肽的抗氧化活性

生物活性肽尝起来很苦且功能特性较差，如低乳化和起泡能力等。因此，该领域研究的方向集中在如何从其他的反应（如美拉德反应）衍生生物活性肽以调节其功能特性并产生一类新的生物活性成分。抗氧化肽（可能具有生物活性）与还原糖，尤其是益生元乳果糖的结合可以产生具有多种用途的高度抗氧化和潜在的营养保健品（如促进健康的功能食品配方）（Nooshkam and Madadlou, 2016a）。

在最近的一项研究中，通过微波加热将乳糖异构化为乳果糖，并通过甲醇法将其纯化为乳果糖含量约 72%的产品（Kareb et al., 2016）。随后，通过微波将乳糖和富含乳果糖的产物（PLu）与乳清蛋白质水解产物（WPH）反应。乳糖表现出比 PLu 更高的美拉德反应性。随着美拉德反应时间的增加，WPH-糖产物对自由基的清除活性增强，美拉德反应也改善了 WPH 的发泡性能。与反应前相比，WPH-糖产物具有更高的溶解度和乳化指数（Nooshkam and Madadlou, 2016b）。这项研究表明，作为一种质优且易于获得的原料，乳清蛋白质在体外被水解成抗氧化肽，然后通过微波触发的美拉德反应与乳糖或乳果糖偶联。此外，在 90℃条件下，PLu 和乳糖皆可与 WPH 发生美拉德反应。对于 WPI-乳糖体系，美拉德反应产物的量高于 WPH-乳糖对应的产物；而 WPH-糖复合物的 DPPH 清除活性明显高于 WPI-糖复合物。基于游离氨基含量的测量结果表明，对于 WPI 和 WPH 的美拉德反应，乳糖比 PLu 更快。傅里叶变换红外光谱证实了乳果糖构型的异头区与 WPH 的结合（Nooshkam and Madadlou, 2016a）。

5.6.4 乳清蛋白质抗氧化活性在食品中的应用

由于成本低廉且抗氧化活性高，诸如丁基羟基茴香醚（BHA）、丁基化羟基甲苯（BHT）、没食子酸丙酯（PG）和叔丁基对苯二酚（TBHQ）等传统的合成抗氧化剂常被用作食品保鲜和防腐的添加剂。但由于其对人体健康的潜在危害，应用又受到限制（Borsato et al., 2014）。现代自由基理论提出，如果氧自由基保持活性,则会导致脂质、蛋白质和 DNA 氧化降解并抑制细胞内抗氧化系统(如氧化损伤)进而诱发糖尿病、心血管疾病、神经组织变性疾病甚至癌症(Tomášková et al.,2014)。因此,寻找一种既能安全抑制脂质氧化又能预防疾病的天然抗氧化剂替代物受到越来越多的关注。

1. 乳清蛋白质抑制脂质氧化

近年来,已经有很多研究聚焦在具有自由基清除和脂质过氧化抑制活性的乳清肽。食物成分的氧化是食物变质的关键因素。众所周知,食品的脂质过氧化会导致食品质量下降、货架期缩短,脂质过氧化产生的自由基可以导致脂肪酸分解,这一过程又会产生不良的味道和有毒物质进而降低食品的营养价值和安全性。因此,在包含脂质和/或脂肪酸的食物中延迟脂质的氧化和自由基的形成显得尤为重要(Brandelli et al.,2015)。

乳清蛋白质可以作为天然抗氧化剂添加到食品中。以下方法常用于提高乳清蛋白质的氧化稳定性:①改变蛋白质结构;②对蛋白质进行基因改造;③将蛋白质作为食品添加剂添加到食品中。将水解的乳清蛋白质添加到脂肪含量高的食品(如肉类产品)中,可减少脂质氧化,从而改善食品质量。例如,将使用复合蛋白酶水解乳清蛋白质得到的 WPH 添加到煮熟的猪肉饼中,可显著抑制脂质氧化中间体共轭二烯和硫代巴比妥酸反应物(TBARS)的生成(Gumus et al.,2017)。而使用胰凝乳蛋白酶和风味蛋白酶仅能延迟共轭二烯的形成,对 TBARS 没有影响。这表明蛋白肽抗氧化活性的提高与蛋白酶的特异性有关。

在最近的一项研究中,通过美拉德反应制备 BSA-右旋糖酐复合物,然后将其作为乳化剂和稳定剂用以生产负载姜黄素的水包油乳状液(Wang et al.,2016)。结果显示,通过热处理形成的乳液在不同温度和 pH 条件下均具有长期稳定性(包括乳液的物理稳定性和负载姜黄素的化学稳定性)。与姜黄素/Tween 20 悬浮液相比,负载姜黄素的 BSA-葡聚糖乳化剂可以使小鼠口服姜黄素的生物利用度提高 4.8 倍。这项研究验证了乳液可以保护负载的姜黄素不被分解,还可以促进姜黄素在胃肠道中的吸收,同时也证明了蛋白多糖乳液是疏水性药物和营养物的良好口服递送系统(Wang et al.,2016)。

开发用于封装和保护化学不稳定的亲脂性食品成分(如富含 omega-3 的油)的输送系统受到越来越多的关注。有学者制备了基于多层乳液的输送系统,该系统由富含 omega-3 的油滴组成,这些油滴被酪蛋白酸盐(Cas)或乳铁蛋白酪蛋白酸盐(Lf-Cas)包被(de Figueiredo Furtado et al.,2016)。含 Lf 和 Cas 的乳液对 pH 变化及盐离子的添加具有更好的物理稳定性。Lf 的添加还能防止乳液中脂质氧化标记物(氢过氧化物和硫代巴比妥酸反应物)的形成,具有增强蛋白质稳定乳剂的物理和化学稳定性的能力,这有利于其用于药物和食品工业的传递系统的开发(Lesmes et al.,2010)。

2. 乳清蛋白质在抗衰老方面的抗氧化作用

在食品加工过程中添加抗氧化肽可以提高食品的稳定性进而延长食品的保存期限。同时，它也可以用作功能性食品、保健食品和抗氧化剂以延缓人体衰老并减少各种疾病的发生。在动物饲料中添加抗氧化肽可以维持动物机体健康或作为饲料防腐剂替代化学抗氧化剂。添加到化妆品中的抗氧化肽在有效抑制皮肤衰老的同时也能抗化妆品氧化，保持其颜色及功能的持久性（Pandey et al., 2018; Brandelli et al., 2015）。

3. 乳清蛋白质在婴儿配方乳粉中的抗氧化作用

开发婴儿营养配方乳粉的重点在于营养成分的组成和功能，包括蛋白质含量和特性（即乳清为主的蛋白质表达谱和α-LA富集）、脂肪酸特性，碳水化合物、维生素以及矿物质的含量与人乳中的含量相匹配。使用 WPH 原料制成的配方乳粉可以根据蛋白质的水解程度进行分类；主要的分类方式是基于氨基酸的分子式，即水解蛋白质/肽后得到的氨基酸。部分水解配方乳粉得到的婴儿营养产品不能以治疗为目的食用，但由于其从配方产品中去除了常见的变应原，因此推荐给有牛奶过敏风险的婴儿食用。部分水解的配方乳粉改善了肠道的消化率和吸收性，因此通常也被称为"预消化"配方粉，食用有助于减少胃肠道不适感（Drapala et al., 2016）。

Lf 作为母乳中天然存在的一种乳蛋白，也常被添加至富含铁的婴儿食品中。最近的一项研究表明，它可以抑制铁催化的氧化（Ueno et al., 2014）。Lf 对婴儿和儿童具有临床益处。大量的临床研究证据也证实了这一点，这些临床研究探索了 Lf 影响婴儿肠道健康和肠道免疫发育及功能的潜在的作用机制（Skarżyńska et al., 2017）。King 及其同事（King et al., 2007）发表了一项随机对照研究，即对妊娠 34 周和 4 周龄喂食配方乳粉的健康婴儿补充 Lf，这些婴儿随机接受添加 Lf（850 mg/L）的配方乳粉或市售配方乳粉（Lf 含 102 mg/L），为期 12 个月，以婴儿出生后第一年的生长变量以及胃肠道、呼吸道和绞痛的发病率为主要衡量指标。结果显示，Lf 强化配方乳粉耐受性良好，且与市售配方乳粉喂养组（0.5 次/年）相比，Lf 强化配方乳粉喂养的婴儿患下呼吸道疾病的概率显著降低（0.15 次/年）。有研究表明，Lf 在促进新生儿和幼儿免疫及防御能力的建立方面发挥主导作用，强有力的试验证据也证实了这一点（Manzoni, 2016）。以上这些研究结果为预防生命早期感染及败血症等相关疾病的发生提供了新思路。目前尚无关于 Lf 治疗的不良反应或对治疗不耐受的报道，因此 Lf 有望在新生儿重症监护病房的感染控制中发挥作用，这些功能作用也有待将来更大规模的试验来证实。

4. 乳清蛋白质在运动饮料中的抗氧化作用

在短跑和耐力运动中，糖原储备的消耗与疲劳有关，因此在运动中保持足够的糖原组织储备是很重要的。与只摄入葡萄糖或水相比，运动前摄入碳水化合物加上WPH会激活骨骼肌中调节运动中葡萄糖摄取和糖原合成的关键酶的蛋白质，从而减轻运动引起的糖原耗竭（Morifuji et al.，2011）。人们对食用乳清蛋白质作为代餐饮料或恢复性运动饮料兴趣浓厚。饮料需要经过热处理以确保安全性和货架稳定性，但是众所周知，乳清蛋白质热稳定性较差，尤其是在静电稳定性减弱的pH或离子强度下。目前，性质稳定的饮料是在酸性或中性条件下配制的，其中主要考虑饮料的涩味和异味。为减少饮料的感官不适，理想的乳清蛋白质饮料pH最好配制为4～6（Wagoner and Foegeding，2017）。为了满足食品药品监督管理局（FDA）对"高"或"优秀"蛋白质来源的要求，每份饮料必须含10 g蛋白质，或每250 mL含约4%蛋白质。满足这些要求和生产标准的乳清蛋白质运动饮料可以激活骨骼肌和调节葡萄糖摄取以最大化运动能力，进而为运动人群提供良好的服务。

5. 乳清蛋白质可食用膜在果蔬保存中的抗氧化作用

食用新鲜、未加工的水果和蔬菜被认为是人类健康的理想选择，因为它们的有效营养成分含量很高。不幸的是，果蔬的易腐性和季节性以及消费者反季节的饮食习惯要求增加了加工和腌制产品的产量。在食物表面涂上可食用的薄膜，可减少水分转移和溶质迁移、气体交换和氧化过程，以及减少甚至抑制可能出现的生理紊乱，从而延长食物的保质期。因此，含有抗菌剂和/或其他食品添加剂（包括抗褐变剂、着色剂、调味剂、营养剂和香料）的可食用涂料正日益成为降低果蔬加工的有害影响以及保持其酚类抗氧化活性的潜在工具。

呼吸作用涉及碳水化合物氧化产生二氧化碳、水和热量的过程。这个过程导致碳水化合物含量减少、质量的减小，还可能导致色泽退化、产生不良气味和味道以及营养价值下降等。可食用薄膜和涂层的阻隔性能通过呼吸速率和耐水蒸气性来评价（Galus and Kadzińska，2015）。蛋白质和复合涂层能对非极性物质（如氧气或二氧化碳）形成巨大屏障。例如，鲜切苹果、土豆和胡萝卜表面涂上乳清蛋白质/果胶混合膜（含转谷氨酰胺酶），分析储存过程中带涂层和未带涂层的果蔬样品的若干特性，交联共混膜的涂层可抑制所有分析样品中微生物的生长，同时还保留了胡萝卜中的酚含量和类胡萝卜素（Marquez et al.，2017）。这些结果表明，用于水果和蔬菜的乳清蛋白质涂层可能对某些食品所含的原始抗氧化剂有保护作用。

5.7 总　　结

乳清蛋白质及其衍生物的抗氧化活性使其具有预防和治疗疾病、保持健康和延缓衰老的潜力。天然存在于食物中的乳清蛋白质可以作为一种有效的抗氧化剂。

改性也可增强乳清蛋白质的抗氧化活性。改性乳清蛋白质对促进健康和治疗疾病的安全性、适口性和耐受性有待进一步研究。乳清蛋白质（或组分）对人类健康产生最优影响的剂量和类型还需要进一步评估。

参 考 文 献

Abreu, S., Santos, R., Moreira, C. et al. (2012). Association between dairy product intake and abdominal obesity in Azorean adolescents. *European Journal of Clinical Nutrition* **66**: 830-835.

Aditya, N.P., Yang, H., Kim, S. et al. (2015). Fabrication of amorphous curcumin nanosuspensions using β-lactoglobulin to enhance solubility, stability, and bioavailability. *Colloids and Surfaces B: Biointerfaces* **127**: 114-121.

Ajello, M., Greco, R., Giansanti, F. et al. (2002). Anti-invasive activity of bovine lactoferrin towards group A streptococci. *Biochemistry and Cell Biology* **80**: 119-124.

Al-Hanish, A., Stanic-Vucinic, D., Mihailovic, J. et al. (2016). Noncovalent interactions of bovine α-lactalbumin with green tea polyphenol, epigalocatechin-3-gallate. *Food Hydrocolloids* **61**: 241-250.

Ameratunga, R. and Woon, S.T. (2010). Anaphylaxis to hyperallergenic functional foods. *Allergy, Asthma & Clinical Immunology* **6**: 33.

Anthony, J.C., Anthony, T.G., Kimball, S.R. et al. (2001). Signaling pathways involved in translational control of protein synthesis in skeletal muscle by leucine. *The Journal of Nutrition* **131** (3): 856S-860S.

Badr, G. (2012). Supplementation with undenatured whey protein during diabetes mellitus improves the healing and closure of diabetic wounds through the rescue of functional long-lived wound macrophages. *Cellular Physiology and Biochemistry* **29**: 571-582.

Badr, G. (2013). Camel whey protein enhances diabetic wound healing in a streptozotocin-induced diabetic mouse model: the critical role of beta-Defensin-1, -2 and -3. *Lipids in Health and Disease* **12**: 46.

Badr, G., Badr, B.M., Mahmoud, M.H. et al. (2012). Treatment of diabetic mice with undenatured whey protein accelerates the wound healing process by enhancing the expression of MIP-1alpha, MIP-2, KC, CX3CL1 and TGF-beta in wounded tissue. *BMC Immunology* **13**: 32.

Balbis, E., Patriarca, S., Furfaro, A. et al. (2009). Whey proteins influence hepatic glutathione after CCl4 intoxication. *Toxicology and Industrial Health* **25**: 325-328.

Bart, P., Yves, B., Senden, J.M.G. et al. (2011). Whey protein stimulates postprandial muscle protein accretion more effectively than do casein and casein hydrolysate in older men. *American Journal of Clinical Nutrition* **93**: 997-1005.

Berkhout, B., Van Wamel, J.L.B., Beljaars, L. et al. (2002). Characterization of the anti-HIV effects of native lactoferrin and other milk proteins and protein-derived peptides. *Antiviral Research* **55**: 341-355.

Bodiga, V.L., Eda, S.R., Veduruvalasa, V.D. et al. (2013). Attenuation of non-enzymatic thermal glycation of bovine serum albumin (BSA) using beta-carotene. *International Journal of*

Biological Macromolecules **56**: 41-48.

Bonang, G., Monintja, H.E., Sujudi, and van der Waaij, D. (2000). Influence of breastmilk on the development of resistance to intestinal colonization in infants born at the Atma Jaya Hospital, Jakarta. *Scandinavian Journal of Infectious Diseases* **32** (2): 189-196.

Bonnefont-Rousselot, D., Ratziu, V., and Giral, P. (2006). Blood oxidative stress markers are unreliable markers of hepatic steatosis. *Alimentary Pharmacology & Therapeutics* **23**: 91-98.

Borsato, D., Cini, J.R.D.M., Silva, H.C.D. et al. (2014). Oxidation kinetics of biodiesel from soybean mixed with synthetic antioxidants BHA, BHT and TBHQ: determination of activation energy. *Fuel Processing Technology* **127**: 111-116.

Botteman, M. and Detzel, P. (2015). Cost-effectiveness of partially hydrolyzed whey protein formula in the primary prevention of atopic dermatitis in high-risk urban infants in Southeast Asia. *Annals of Nutrition and Metabolism* **66** (S1): 26-32.

Bounous, G. (2000). Whey protein concentrate (WPC) and glutathione modulation in cancer treatment. *Anticancer Research* **20**: 4785-4792.

Bounous, G., Batist, G., and Gold, P. (1991). Whey proteins in cancer prevention. *Cancer Letters* **57**: 91-94.

Bounous, G., Baruchel, S., Falutz, J. et al. (1993). Whey proteins as a food supplement in HIVseropositive individuals. *Clinical & Investigative Medicine Médecine Clinique Et Experimentale* **16** (3): 204-209.

Brandelli, A., Daroit, D.J., and Corrêa, A.P.F. (2015). Whey as a source of peptides with remarkable biological activities. *Food Research International* **73**: 149-161.

Bu, G.H., Luo, Y.K., Zheng, Z. et al. (2009). Effect of heat treatment on the antigenicity of bovine α-lactalbumin and β-lactoglobulin in whey protein isolate. *Food and Agricultural Immunology* **20** (3): 195-206.

de Castro, R.J.S. and Sato, H.H. (2015). Biologically active peptides: processes for their generation, purification and identification and applications as natural additives in the food and pharmaceutical industries. *Food Research International* **74**: 185-198.

Chatterton, D.E.W., Smithers, G., Roupas, P. et al. (2006). Bioactivity of β-lactoglobulin and α-lactalbumin-technological implications for processing. *International Dairy Journal* **16**: 1229-1240.

Chawla, S.P., Chander, R., and Sharma, A. (2009). Antioxidant properties of Maillard reaction products obtained by gamma-irradiation of whey proteins. *Food Chemistry* **116**: 122-128.

Chitapanarux, T., Tienboon, P., Pojchamarnwiputh, S. et al. (2009). Open-labeled pilot study of cysteine-rich whey protein isolate supplementation for nonalcoholic steatohepatitis patients. *Journal of Gastroenterology and Hepatology* **24** (6): 1045-1050.

Chungchunlam, S.M.S., Henare, S.J., Ganesh, S. et al. (2014). Effect of whey protein and glycomacropeptide on measures of satiety in normal-weight adult women. *Appetite* **78**: 172-178.

Coburn, J.W., Housh, D.J., Housh, T.J. et al. (2006). Effects of leucine and whey protein supplementation during eight weeks of unilateral resistance training. *The Journal of Strength & Conditioning Research* **20**: 284-291.

Cvander, V., Gruppen, H., Dbade, B. et al. (2002). Optimisation of the angiotensin converting enzyme inhibition by whey protein hydrolysates using response surface methodology. *International Dairy Journal* **12**: 813-820.

Da Silva, S.V., Picolotto, R.S., Wagner, R. et al. (2015). Elemental (macro-and microelements) and amino acid profile of milk proteins commercialized in Brazil and their nutritional value. *Journal of Food and Nutrition Research* **3** (7): 430-436.

Daenzer, M., Petzke, K.J., Bequette, B.J. et al. (2001). Whole-body nitrogen and splanchnic amino acid metabolism differ in rats fed mixed diets containing casein or its corresponding amino acid mixture. *The Journal of Nutrition* **131** (7): 1965-1972.

Di Mario, F., Aragona, G., Dal Bò, N. et al. (2003). Use of bovine lactoferrin for helicobacter pylori eradication. *Digestive and Liver Disease* **35** (10): 706-710.

Drapala, K.P., Auty, M.A.E., Mulvihill, D.M. et al. (2016). Improving thermal stability of hydrolysed whey protein-based infant formula emulsions by protein-carbohydrate conjugation. *Food Research International* **88**: 42-51.

Dupont, C. (2003). Protein requirements during the first year of life. *The American Journal of Clinical Nutrition* **77** (6): 1544S-1549S.

Ebaid, H., Salem, A., Sayed, A. et al. (2011). Whey protein enhances normal inflammatory responses during cutaneous wound healing in diabetic rats. *Lipids in Health and Disease* **10**: 235.

Ebaid, H., Ahmed, O.M., Mahmoud, A.M. et al. (2013). Limiting prolonged inflammation during proliferation and remodeling phases of wound healing in streptozotocin-induced diabetic rats supplemented with camel undenatured whey protein. *BMC immunology* **14**: 31.

El-Desouky, W.I., Mahmoud, A.H., and Abbas, M.M. (2017). Antioxidant potential and hypolipidemic effect of whey protein against gamma irradiation induced damages in rats. *Applied Radiation and Isotopes* **129**: 103-107.

Exl, B. (2001). A review of recent developments in the use of moderately hydrolyzed whey formulae in infant nutrition. *Nutrition Research* **21** (1-2): 355-379.

Fan, Y., Yi, J., Zhang, Y. et al. (2018). Fabrication of curcumin-loaded bovine serum albumin (BSA)-dextran nanoparticles and the cellular antioxidant activity. *Food Chemistry* **239**: 1210-1218.

Farnfield, M.M., Carey, K.A., Gran, P. et al. (2009a). Whey protein ingestion activates mTORdependent signalling after resistance exercise in young men: a double-blinded randomized controlled trial. *Nutrients* **1**: 263-275.

Farnfield, M.M., Trenery, C., Carey, K.A. et al. (2009b). Plasma amino acid response after ingestion of different whey protein fractions. *International Journal of Food Sciences and Nutrition* **60**: 476-486.

de Figueiredo Furtado, G., Michelon, M., de Oliveira, D.R.B. et al. (2016). Heteroaggregation of lipid droplets coated with sodium caseinate and lactoferrin. *Food Research International* **89**: 309-319.

Fonseca, D.P., Khalil, N.M., and Mainardes, R.M. (2017). Bovine serum albumin-based nanoparticles containing resveratrol: characterization and antioxidant activity. *Journal of Drug Delivery Science and Technology* **39**: 147-155.

Galus, S. and Kadzińska, J. (2015). Food applications of emulsion-based edible films and coatings.

Trends in Food Science and Technology **45** (2): 273-283.

Golzar, F.A.K., Vatani, D.S., Mojtahedi, H. et al. (2012). The effects of whey protein isolate supplementation and resistance training on cardiovascular risk factors in overweight young men. *Journal of Isfahan Medical School* **30** (181).

Gregory, K.E. and Walker, W.A. (2013). Immunologic factors in human milk and disease prevention in the preterm infant. *Current Pediatrics Reports* **1** (4): 222-228.

Grimble, R.F. (2006). The effects of sulfur amino acid intake on immune function in humans. *The Journal of Nutrition* **136** (6): 1660S-1665S.

Gu, Y. and Wu, J. (2016). Bovine lactoferrin-derived ACE inhibitory tripeptide LRP also shows antioxidative and anti-inflammatory activities in endothelial cells. *Journal of Functional Foods* **25**: 375-384.

Gumus, C.E., Decker, E.A., and McClements, D.J. (2017). Impact of legume protein type and location on lipid oxidation in fish oil-in-water emulsions: lentil, pea, and faba bean proteins. *Food Research International* **100**: 175-185.

Gunnerud, U., Holst, J.J., Ostman, E. et al. (2012). The glycemic, insulinemic and plasma amino acid responses to equi-carbohydrate milk meals, a pilot- study of bovine and human milk. *Nutrition Journal* **11**: 83.

Ha, E. and Zemel, M.B. (2003). Functional properties of whey, whey components, and essential amino acids: mechanisms underlying health benefits for active people (review). *The Journal of Nutritional Biochemistry* **14**: 251-258.

Hakkak, R., Korourian, S., Shelnutt, S.R. et al. (2000). Diets containing whey proteins or soy protein isolate protect against 7,12-dimethylbenz(a)anthracene-induced mammary tumors in female rats. *Cancer Epidemiology Biomarkers & Prevention* **9**: 113-117.

Hakkak, R., Korourian, S., Ronis, M. et al. (2001). Dietary whey protein protects against azoxymethane-induced colon tumors in male rats. *Cancer Epidemiology Biomarkers & Prevention* **10** (5): 555-558.

Hansen, K.B., Vilsbøll, T., Bagger, J.I. et al. (2010). Reduced glucose tolerance and insulin resistance induced by steroid treatment, relative physical inactivity, and high-calorie diet impairs the incretin effect in healthy subjects. *Journal of Clinical Endocrinology & Metabolism* **95** (7): 3309-3317.

Harper, A.E., Miller, R.H., and Block, K.P. (1984). Branched-chain amino acid metabolism. *Annual Review of Nutrition* **4**: 409-454.

Hernándezledesma, B., Miguel, M., and Amigo, L. (2007). Effect of simulated gastrointestinal digestion on the antihypertensive properties of synthetic β-lactoglobulin peptide sequences. *Journal of Dairy Research* **74** (3): 336-339.

Hochwallner, H. et al. (2014). Cow's milk allergy: from allergens to new forms of diagnosis, therapy and prevention. *Methods* **66** (1): 22-33.

Hoefle, A.S., Bangert, A.M., Stamfort, A. et al. (2015). Metabolic responses of healthy or prediabetic adults to bovine whey protein and sodium caseinate do not differ. *Journal of Nutrition* **145** (3): 467-475.

Hongsprabhas, P., Kerdchouay, P., and Sakulsom, P. (2011). Lowering the Maillard reaction products

(MRPs) in heated whey protein products and their cytotoxicity in human cell models by whey protein hydrolysate. *Food Research International* **44** (3): 748-754.

Høst, A. (2002). Frequency of cow's milk allergy in childhood. *Annals of Allergy Asthma & Immunology* **89** (1): 33-37.

Ikeda, M., Sugiyama, K., Tanaka, T. et al. (1998). Lactoferrin markedly inhibits hepatitis C virus infection in cultured human hepatocytes. *Biochemical & Biophysical Research Communications* **245** (2): 549-553.

Jiang, Z. and Brodkorb, A. (2012). Structure and antioxidant activity of Maillard reaction products from α-lactalbumin and β-lactoglobulin with ribose in an aqueous model system. *Food Chemistry* **133** (1): 960-968.

Jiménez-Castaño, L., Villamiel, M., and López-Fandiño, R. (2007). Glycosylation of individual whey proteins by Maillard reaction using dextran of different molecular mass. *Food Hydrocolloids* **21** (3): 433-443.

Joubran, Y., Mackie, A., and Lesmes, U. (2013). Impact of the Maillard reaction on the antioxidant capacity of bovine lactoferrin. *Food Chemistry* **141** (4): 3796-3802.

Kareb, O., Champagne, C.P., and Aider, M. (2016). Contribution to the production of lactulose-rich whey by in situ electro-isomerization of lactose and effect on whey proteins after electro-activation as confirmed by matrix-assisted laser desorption/ionization time-of-flight-mass spectrometry and sodium dodecyl sulfate-polyacrylamide gel electrophoresis. *Journal of Dairy Science* **99**: 2552-2570.

Karelis, A.D., Messier, V., Suppere, C. et al. (2015). Effect of cysteine-rich whey protein (immunocal(R)) supplementation in combination with resistance training on muscle strength and lean body mass in non-frail elderly subjects: a randomized, double-blind controlled study. *Journal of Nutrition Health & Aging* **19**: 531-536.

Katsanos, C.S., Kobayashi, H., Sheffieldmoore, M. et al. (2006). A high proportion of leucine is required for optimal stimulation of the rate of muscle protein synthesis by essential amino acids in the elderly. *American Journal of Physiology Endocrinology and Metabolism* **291** (2): E381-E387.

Kattan, J.D., Cocco, R.R., and Jarvinen, K.M. (2011). Milk and soy allergy. *Pediatric Clinics of North America* **58** (2): 407-426.

Kawase, M. (2000). Effect of administration of fermented milk containing whey protein concentrate to rats and healthy men on serum lipids and blood pressure. *Journal of Dairy Science* **83** (2): 255-263.

Kennedy, R.S., Konok, G.P., Bounous, G. et al. (1995). The use of a whey protein concentrate in the treatment of patients with metastatic carcinoma: a phase I-II clinical study. *Anticancer Research* **15**: 2643-2649.

Khan, R.S., Grigor, J., Winger, R. et al. (2013). Functional food product development- opportunities and challenges for food manufacturers. *Trends in Food Science & Technology* **30** (1): 27-37.

King, J.C.J., Cummings, G.E., Guo, N. et al. (2007). A double-blind, placebo-controlled, pilot study of bovine lactoferrin supplementation in bottle-fed infants. *Journal of Pediatric Gastroenterology and Nutrition* **44** (2): 245-251.

Kokoszka, S., Debeaufort, F., Lenart, A. et al. (2010). Liquid and vapour water transfer through whey protein/lipid emulsion films. *Journal of the Science of Food and Agriculture* **90**: 1673-1680.

Kulczycki, J.A. and Macdermott, R.P. (1985). Bovine IgG and human immune responses: Con Ainduced mitogenesis of human mononuclear cells is suppressed by bovine IgG. *International Archives of Allergy & Applied Immunology* **77** (1-2): 255-258.

Laursen, I., Briand, P., and Lykkesfeldt, A.E. (1989). Serum albumin as a modulator on growth of the human breast cancer cell line MCF-7. *Anticancer Research* **10** (2A): 343-351.

Lavie, C.J., Milani, R.V., and Ventura, H.O. (2009). Obesity and cardiovascular disease: risk factor, paradox, and impact of weight loss. *Journal of the American College of Cardiology* **53** (21): 1925-1932.

Lesmes, U., Sandra, S., Decker, E.A. et al. (2010). Impact of surface deposition of lactoferrin on physical and chemical stability of omega-3 rich lipid droplets stabilised by caseinate. *Food Chemistry* **123** (1): 99-106.

Li, M., Ma, Y., and Ngadi, M.O. (2013). Binding of curcumin to β-lactoglobulin and its effect on antioxidant characteristics of curcumin. *Food Chemistry* **141** (2): 1504-1511.

Liu, Q., Kong, B., Han, J. et al. (2014). Structure and antioxidant activity of whey protein isolate conjugated with glucose via the Maillard reaction under dry-heating conditions. *Food Structure* **1** (2): 145-154.

Liu, F., Wang, D., Sun, C. et al. (2016a). Influence of polysaccharides on the physicochemical properties of lactoferrin-polyphenol conjugates coated β-carotene emulsions. *Food Hydrocolloids* **52**: 661-669.

Liu, F., Wang, D., Xu, H. et al. (2016b). Physicochemical properties of β-carotene emulsions stabilized by chlorogenic acid-lactoferrin-glucose/polydextrose conjugates. *Food Chemistry* **196**: 338-346.

Losso, J.N., Dhar, J., Kummer, A. et al. (1993). Detection of antibody specificity of raw bovine and human milk to bacterial lipopolysaccharides using PCFIA. *Food & Agricultural Immunology* **5** (4): 231-239.

Lykholat, O.A., Grigoryuk, I.P., and Lykholat, T.Y. (2016). Metabolic effects of alimentary estrogen in different age animals. *Annals of Agrarian Science* **14** (4): 335-339.

Manzoni, P. (2016). Clinical benefits of lactoferrin for infants and children. *The Journal of Pediatrics* **173**: S43-S52.

Marquez, R.G., Pierro, D.P., Mariniello, L. et al. (2017). Fresh-cut fruit and vegetable coatings by transglutaminase-crosslinked whey protein/pectin edible films. *LWT-Food Science and Technology* **75**: 124-130.

Marshall, K. (2004). Therapeutic applications of whey protein. *Alternative Medicine Review* **9**: 136-156.

Marsset-Baglieri, A., Fromentin, G., Airinei, G. et al. (2014). Milk protein fractions moderately extend the duration of satiety compared with carbohydrates independently of their digestive kinetics in overweight subjects. *British Journal of Nutrition* **112**: 557-564.

Martínez-Maqueda, D., Miralles, B., Ramos, M. et al. (2013). Effect of β-lactoglobulin hydrolysate and β-lactorphin on intestinal mucin secretion and gene expression in human goblet cells. *Food Research International* **54** (1): 1287-1291.

Mcdonough, F.E., Hargrove, R.E., Mattingly, W.A. et al. (1974). Composition and properties of whey protein concentrates from ultrafiltration. *Journal of Dairy Science* **57** (12): 1438-1443.

Micke, P., Beeh, K.M., Schlaak, J.F. et al. (2001). Oral supplementation with whey proteins increases plasma glutathione levels of HIV-infected patients. *European Journal of Clinical Investigation* **31** (2): 171-178.

Mikael, N., Holst, J.J., and Ingerme, B.R. (2007). Metabolic effects of amino acid mixtures and whey protein in healthy subjects: studies using glucose-equivalent drinks. *American Journal of Clinical Nutrition* **85**: 996-1004.

Millward, D.J., Layman, D.K., Tome, D. et al. (2008). Protein quality assessment: impact of expanding understanding of protein and amino acid needs for optimal health. *The American Journal of Clinical Nutrition* **87** (5): 1576S-1581S.

Mobley, C.B., Fox, C.D., Thompson, R.M. et al. (2016). Comparative effects of whey protein versus lleucine on skeletal muscle protein synthesis and markers of ribosome biogenesis following resistance exercise. *Amino Acids* **48**: 733-750.

Morifuji, M., Kanda, A., Koga, J. et al. (2011). Preexercise ingestion of carbohydrate plus whey protein hydrolysates attenuates skeletal muscle glycogen depletion during exercise in rats. *Nutrition* **27** (7-8): 833-837.

Naso, L.G., Lezama, L., ValcarceL, M. et al. (2016). Bovine serum albumin binding, antioxidant and anticancer properties of an oxidovanadium(IV) complex with luteolin. *Journal of Inorganic Biochemistry* **157**: 80-93.

Nilsson, M. (2006). Insulinogenic Effects of Milk- and Other Dietary Proteins, Mechanisms and metabolic implications. Doctoral thesis. Lund University.

Nilsson, M., Holst, J.J., and Björck, I.M. (2007). Metabolic effects of amino acid mixtures and whey protein in healthy subjects: studies using glucose-equivalent drinks. *The American Journal of Clinical Nutrition* **85** (4): 996-1004.

Nooshkam, M. and Madadlou, A. (2016a). Maillard conjugation of lactulose with potentially bioactive peptides. *Food Chemistry* **192**: 831-836.

Nooshkam, M. and Madadlou, A. (2016b). Microwave-assisted isomerisation of lactose to lactulose and Maillard conjugation of lactulose and lactose with whey proteins and peptides. *Food Chemistry* **200**: 1-9.

O'Loughlin, I.B., Murray, B.A., Brodkorb, A. et al. (2014). Production of whey protein isolate hydrolysate fractions with enriched ACE-inhibitory activity. *International Dairy Journal* **38** (2): 101-103.

Paddon-Jones, D., Campbell, W.W., Jacques, P.F. et al. (2015). Protein and healthy aging. *American Journal of Clinical Nutrition* **101** (6): https://doi.org/10.3945/ajcn.114.084061.

Pal, S., Ellis, V., and Ho, S. (2010). Acute effects of whey protein isolate on cardiovascular risk factors in overweight, post-menopausal women. *Atherosclerosis* **212** (1): 339-344.

Pan, D., Cao, J., Guo, H. et al. (2012). Studies on purification and the molecular mechanism of a novel ACE inhibitory peptide from whey protein hydrolysate. *Food Chemistry* **130**: 121-126.

Panahi, S. (2014). Milk and its components in the regulation of short-term appetite, food intake and glycemia in young adults. Doctoral thesis. University of Toronto.

Pandey, M., Kapila, R., and Kapila, S. (2018). Osteoanabolic activity of whey-derived anti-oxidative (MHIRL and YVEEL) and angiotensin-converting enzyme inhibitory (YLLF, ALPMHIR, IPA and WLAHK) bioactive peptides. *Peptides* **99**: 1-7.

Peña-Ramos, E.A. and Xiong, Y.L. (2001). Antioxidative activity of whey protein hydrolysates in a liposomal system. *Journal of Dairy Science* **84** (12): 2577-2583.

Pins, J.J. and Keenan, J.M. (2006). Effects of whey peptides on cardiovascular disease risk factors. *Journal of Clinical Hypertension* **8** (11): 775-782.

Prashanth, M.K., Madaiah, M., Revanasiddappa, H.D. et al. (2013). Synthesis, anticonvulsant, antioxidant and binding interaction of novel N-substituted methylquinazoline-2,4(1H, 3H)-dione derivatives to bovine serum albumin: a structure-activity relationship study. *Spectrochimica Acta Part A: Molecular and Biomolecular Spectroscopy* **110** (6): 324-332.

Rani, A.J. and Mythili, S.V. (2014). Study on total antioxidant status in relation to oxidative stress in type 2 diabetes mellitus. *Journal of Clinical & Diagnostic Research* **8** (3): 108-110.

Requena, P., Lopez-Posadas, R., Abadia-Molina, A. et al. (2009). M1663 The Intestinal Antiinflammatory Agent Glycomacropeptide Has Immunomodulatory Effects on Rat. Splenocytes. *Gastroenterology* **136**: A-405.

Requena, P., González, R., López-Posadas, R. et al. (2010). The intestinal antiinflammatory agent glycomacropeptide has immunomodulatory actions on rat splenocytes. *Biochemical Pharmacology* **79** (12): 1797-1804.

Rocha, G.F., Kise, F., Rosso, A.M. et al. (2017). Potential antioxidant peptides produced from whey hydrolysis with an immobilized aspartic protease from Salpichroa originifolia fruits. *Food Chemistry* **237**: 350-355.

Rusu, D., Drouin, R., Pouliot, Y. et al. (2009). A bovine whey protein extract can enhance innate immunity by priming normal human blood neutrophils. *The Journal of Nutrition* **139** (2): 386-393.

Sadowska-Bartosz, I., Stefaniuk, I., Galiniak, S. et al. (2015). Glycation of bovine serum albumin by ascorbate in vitro: possible contribution of the ascorbyl radical? *Redox Biology* **6**: 93-99.

Sah, B.N.P., Mcainch, A.J., and Vasiljevic, T. (2016). Modulation of bovine whey protein digestion in gastrointestinal tract: a comprehensive review. *International Dairy Journal* **62**: 10-18.

Sarwar, G. (1997). The protein digestibility-corrected amino acid score method overestimates quality of proteins containing antinutritional factors and of poorly digestible proteins supplemented with limiting amino acids in rats. *Journal of Nutrition* **127**: 758-764.

Schaafsma, G. (2000). The protein digestibility-corrected amino acid score. *The Journal of Nutrition* **130** (7): 1865S-1867S.

Schröder, A., Berton-Carabin, C., Venema, P. et al. (2017). Interfacial properties of whey protein and whey protein hydrolysates and their influence on O/W emulsion stability. *Food Hydrocolloids* **73**: 129-140.

See, D., Mason, S., and Roshan, R. (2002). Increased tumor necrosis factor alpha (Tnf-) and natural killer cell (NK) function using an integrative approach in late stage cancer. *Immunological Investigations* **31** (2): 137-153.

Sekine, K., Watanabe, E., Nakamura, J. et al. (1997). Inhibition of azoxymethane-initiated colon tumor by bovine lactoferrin administration in F344 rats. *Japanese Journal of Cancer Research* **88**: 523-526.

Shimomura, Y., Murakami, T., Nakai, N. et al. (2004). Exercise promotes BCAA catabolism: effects of BCAA supplementation on skeletal muscle during exercise. *The Journal of Nutrition* **134** (6): 1583S-1587S.

Shute, M. (2004). Effect of whey protein isolate on oxidative stress, exercise performance, and immunity. Doctoral Dissertation. Virginia Polytechnic and State University.

Sigal, R.J., Kenny, G.P., Wasserman, D.H. et al. (2006). Physical activity/exercise and type 2 diabetes a consensus statement from the American Diabetes Association. *Diabetes Care* **29** (6): 1433-1438.

Sila, A. and Bougatef, A. (2016). Antioxidant peptides from marine by-products: isolation, identification and application in food systems. A review. *Journal of Functional Foods* **21**: 10-26.

Skarżyńska, E., Żytyńska-Daniluk, J., and Lisowska-Myjak, B. (2017). Correlations between ceruloplasmin, lactoferrin and myeloperoxidase in meconium. *Journal of Trace Elements in Medicine and Biology* **43**: 58-62.

Smithers, G.W., Johnson, M.A., Jelen, P. et al. (1998). Anti-cancer effects of dietary whey proteins. In: *Proceedings of the Second International Whey Conference*, Chicago, USA (27-29 October 1997). Brussels, Belgium: International Dairy Federation.

Song, W., Tavitian, A., Cressatti, M. et al. (2017). Cysteine-rich whey protein isolate (Immunocal®) ameliorates deficits in the GFAP.HMOX1 mouse model of schizophrenia. *Free Radical Biology and Medicine* **110**: 162-175.

Stanic-Vucinic, D., Prodic, I., Apostolovic, D. et al. (2013). Structure and antioxidant activity of β-lactoglobulin-glycoconjugates obtained by high-intensity-ultrasound-induced Maillard reaction in aqueous model systems under neutral conditions. *Food Chemistry* **138** (1): 590-599.

Stobaugh, H.C., Ryan, K.N., Kennedy, J.A. et al. (2016). Including whey protein and whey permeate in ready-to-use supplementary food improves recovery rates in children with moderate acute malnutrition: a randomized, double-blind clinical trial. *The American Journal of Clinical Nutrition* **103** (3): 926-933.

Stojadinovic, M., Radosavljevic, J., Ognjenovic, J. et al. (2013). Binding affinity between dietary polyphenols and β-lactoglobulin negatively correlates with the protein susceptibility to digestion and total antioxidant activity of complexes formed. *Food Chemistry* **136** (3-4): 1263-1271.

Sun, Y., Hayakawa, S., Puangmanee, S. et al. (2006). Chemical properties and antioxidative activity of glycated α-lactalbumin with a rare sugar, d-allose, by Maillard reaction. *Food Chemistry* **95** (3): 509-517.

Tahavorgar, A., Vafa, M., Shidfar, F. et al. (2015). Beneficial effects of whey protein preloads on some cardiovascular diseases risk factors of overweight and obese men are stronger than soy protein preloads—A randomized clinical trial. *Journal of Nutrition & Intermediary Metabolism* **2**: 69-75.

Tanaka, K., Ikeda, M., Nozaki, A. et al. (1999). Lactoferrin inhibits hepatitis C virus viremia in patients with chronic hepatitis C: a pilot study. *Cancer Science* **90**: 367-371.

Tang, J.E., Moore, D.R., Kujbida, G.W. et al. (2009). Ingestion of whey hydrolysate, casein, or soy protein isolate: effects on mixed muscle protein synthesis at rest and following resistance exercise in young men. *Journal of Applied Physiology* **107** (3): 987-992.

Tatpati, L., Bigelow, M.L., Irving, B.A. et al. (2007). Branched chain amino acids effect on skeletal muscle mitochondrial ATP production and protein metabolism-interaction with insulin and age. *Diabetes* **56**: 343-348.

Tina, A., Luhovyy, B.L., Brown, P.H. et al. (2010). Effect of premeal consumption of whey protein and its hydrolysate on food intake and postmeal glycemia and insulin responses in young adults. *The American Journal of Clinical Nutrition* **91** (4): 966-975.

Tipton, K.D., Elliott, T.A., Cree, M.G. et al. (2004). Ingestion of casein and whey proteins result in muscle anabolism after resistance exercise. *Medicine and Science in Sports and Exercise* **36**: 2073-2081.

Tomášková, M., Chýlková, J., Jehlička, V.R. et al. (2014). Simultaneous determination of BHT and BHA in mineral and synthetic oils using linear scan voltammetry with a gold disc electrode. *Fuel* **123**: 107-112.

Tsuda, H., Sekine, K., Nakamura, J. et al. (1998). Inhibition of azoxymethane initiated colon tumor and aberrant crypt foci development by bovine lactoferrin administration in F344 rats. *Advances in Experimental Medicine & Biology* **443**: 273-284.

U.S. Dairy Export Council. (1999). Reference Manual for U.S. Whey Products. 2nd Edition, U.S.D.E. Council, Editor.

Ueno, H.M., Urazono, H., and Kobayashi, T. (2014). Serum albumin forms a lactoferrin-like soluble iron-binding complex in presence of hydrogen carbonate ions. *Food Chemistry* **145**: 90-94.

Vavrusova, M., Pindstrup, H., Johansen, L.B. et al. (2015). Characterisation of a whey protein hydrolysate as antioxidant. *International Dairy Journal* **47**: 86-93.

Wagoner, T.B. and Foegeding, E.A. (2017). Whey protein-pectin soluble complexes for beverage applications. *Food Hydrocolloids* **63**: 130-138.

Wang, H., Ye, X., and Ng, T.B. (2000). First demonstration of an inhibitory activity of milk proteins against human immunodeficiency virus-1 reverse transcriptase and the effect of succinylation. *Life Sciences* **67** (22): 2745-2752.

Wang, Y.Z., Xu, C.L., An, Z.H. et al. (2008). Effect of dietary bovine lactoferrin on performance and antioxidant status of piglets. *Animal Feed Science and Technology* **140** (3-4): 326-336.

Wang, W.Q., Bao, Y.H., and Chen, Y. (2013). Characteristics and antioxidant activity of water-soluble Maillard reaction products from interactions in a whey protein isolate and sugars system. *Food Chemistry* **139** (1-4): 355-361.

Wang, C., Liu, Z., Xu, G. et al. (2016). BSA-dextran emulsion for protection and oral delivery of curcumin. *Food Hydrocolloids* **61**: 11-19.

Watanabe, A., Okada, K., Shimizu, Y. et al. (2000). Nutritional therapy of chronic hepatitis by whey protein (non-heated). *Journal of Medicine* **31** (5-6): 283-302.

Weinberg, E.D. (1996). The role of iron in cancer. *European Journal of Cancer Prevention* **5**: 19-36.

Wildová, E. and Anděl, M. (2013). Casein and whey proteins are the physiological stimuli of the insulin secretion. *Diabetologie Metabolismus Endokrinologie Vyziva* **16** (3): 179-185.

Wu, G., Fang, Y.Z., Yang, S. et al. (2004). Glutathione metabolism and its implications for health. *The Journal of Nutrition* **134** (3): 489-492.

Wu, S., Zhang, Y., Ren, F. et al. (2018). Structure-affinity relationship of the interaction between phenolic acids and their derivatives and β-lactoglobulin and effect on antioxidant activity. *Food Chemistry* **245**: 613-619.

Yadav, J.S.S., Yan, S., Pilli, S. et al. (2015). Cheese whey: a potential resource to transform into bioprotein, functional/nutritional proteins and bioactive peptides. *Biotechnology Advances* **33** (6): 756-774.

Yao, X.H., Zhang, Z.B., Song, P. et al. (2016). Different harvest seasons modify bioactive compounds and antioxidant activities of Pyrola incarnata. *Industrial Crops and Products* **94**: 405-412.

Yi, J., Fan, Y., Zhang, Y. et al. (2016). Glycosylated α-lactalbumin-based nanocomplex for curcumin: physicochemical stability and DPPH-scavenging activity. *Food Hydrocolloids* **61**: 369-377.

Yolken, R.H., Losonsky, G.A., Vonderfecht, S. et al. (1985). Antibody to human rotavirus in cow's milk. *New England Journal of Medicine* **312**: 605-610.

Yoo, Y.C., Watanabe, S., Watanabe, R. et al. (1997). Bovine lactoferrin and lactoferricin, a peptide derived from bovine lactoferrin, inhibit tumor metastasis in mice. *Japanese Journal of Cancer Research* **88** (2): 184-190.

Zemel, M.B. and Zhao, F. (2009). Role of whey protein and whey components in weight management and energy metabolism. *Journal of Hygiene Research* **38**: 114-117.

Zhang, T., Jiang, B., Miao, M. et al. (2012). Combined effects of high-pressure and enzymatic treatments on the hydrolysis of chickpea protein isolates and antioxidant activity of the hydrolysates. *Food Chemistry* **135**: 904-912.

Zhao, F.X. (2010). Protective effects of bioactive peptides of whey protein on cardiovascular health. *Food & Nutrition in China* **2**: 74-77.

Zheng, G., Liu, H., Zhu, Z. et al. (2016a). Selenium modification of β-lactoglobulin (β-Lg) and its biological activity. *Food Chemistry* **204**: 246-251.

Zheng, G., Xu, X., Zheng, J. et al. (2016b). Protective effect of seleno-β-lactoglobulin (Se-β-lg) against oxidative stress in D-galactose-induced aging mice. *Journal of Functional Foods* **27**: 310-318.

第 6 章 乳清蛋白质在营养产品中的应用

Mingruo Guo[1,2] and Guorong Wang[1]

1. Department of Nutrition and Food Science, University of Vermont, Burlington, USA
2. College of Food Science, Northeast Agriculture University, Harbin, People's Republic of China

作为一种优质经济的蛋白质，乳清蛋白质因其具有高品质氨基酸组成（Hoffman and Falvo，2004）、风味清爽（Evans et al.，2009）等特点得到广泛应用。乳清蛋白质能够在较低的 pH 条件下溶解，同时可以添加到不同的食品当中（图 6.1）。人乳不同于牛乳，相对于酪蛋白来说，人乳含有更多的乳清蛋白质。因此，乳清蛋白质与酪蛋白的比例自婴儿配方奶粉问世以来便一直存在。乳清蛋白质含有丰富的支链氨基酸（BCAA）和必需氨基酸（EAA），具有促进肌肉组织合成和防止肌肉组织大量流失等特定功能。而这些特性使乳清蛋白质逐渐成为运动员、老年人和任何想增肌或预防肌肉减少人群的理想蛋白质补充物（Breen and Phillips，2011；de Aguilar-Nascimento et al.，2011）。本章将介绍乳清蛋白质在几种主要营养食品配方中的应用。

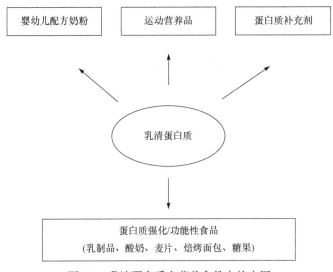

图 6.1 乳清蛋白质在营养食品中的应用

6.1 婴幼儿配方奶粉

6.1.1 乳清蛋白质与酪蛋白比例

母乳是婴幼儿的理想食物。研究表明，母乳喂养对婴幼儿和母亲都有很多益处（Lawrence，2000；Hendricks and Guo，2014）。母乳含有新生儿所需的大部分营养、能量，以及抗体、酶、生长因子、胃肠保护因子和免疫细胞（Newburg，2001；Hendricks and Guo，2014）。然而，出于各种原因，母乳喂养可能受到部分限制（Ogbuanu et al.，2009）。

由于母乳和牛乳的成分不同，牛乳不能直接喂给婴幼儿（Guo，2014）。牛乳和人乳在常量营养素方面的差异列于表6.1。

表6.1 牛乳和人乳常量营养素含量及组成

	人乳	牛乳
总蛋白质（%）	1.00	3.40
乳清蛋白质与酪蛋白比例	60∶40	20∶80
酪蛋白（%）	0.3	2.7
α_{s1}-酪蛋白（%）（酪蛋白总量）	—	40
α_{s2}-酪蛋白（%）（酪蛋白总量）	—	8
β-酪蛋白（%）（酪蛋白总量）	85	38
κ-酪蛋白（%）（酪蛋白总量）	15	12
乳清蛋白质，%	0.7	0.7
α-乳清蛋白质（%）（乳清蛋白质总量）	26	17
β-乳球蛋白（%）（乳清蛋白质总量）	—	43
乳铁蛋白（%）（乳清蛋白质总量）	26	痕量
血清白蛋白（%）（乳清蛋白质总量）	10	5
溶菌酶（%）（乳清蛋白质总量）	10	痕量
免疫球蛋白（%）（乳清蛋白质总量）	16（IgA）	10（IgG）
脂肪（%）	3.80	3.50
乳糖（%）	7.00	5.00
灰分（%）	0.20	0.70

注：资料改编自Guo（2014）。

乳清蛋白质与酪蛋白的比例则是区分二者的重要条件之一。在牛乳中，乳清蛋白质与酪蛋白的比例通常是20∶80。在母乳中，这个比例在哺乳早期是90∶10，在哺乳中期是60∶40，在哺乳后期是50∶50（Kunz and Lonnerdal，1992）。在行业内以60∶40比例为标准。在婴幼儿配方奶粉生产中，乳清蛋白质（脱盐乳清粉、WPC和WPI）通常被添加到牛乳或奶粉中，目的是调节乳清蛋白质与酪蛋白的比例。例如，如果用牛乳作为婴幼儿配方奶粉的基础原料，在100 g牛乳中需加入约3.35 g乳清蛋白质，使得乳清蛋白质与酪蛋白的比例从20∶80调整到60∶40。然后，通过添加植物油和乳糖来调整脂肪和乳糖含量。在婴幼儿配方奶粉的生产过程中，这样的配比是调节牛乳成分使其更接近母乳的常用方法。婴幼儿配方奶粉中的总蛋白质含量可能略高于母乳，这是由于配方奶粉与母乳中的氨基酸含量并不完全匹配。出现少量过剩现象，则可能是为了确保婴儿通过奶瓶喂养能够获得足够的氨基酸，而婴幼儿配方奶粉中，较低的蛋白质摄入可能与两岁以下婴儿体重较低有关（European Childhood Obesity Trial Study Group，2009）。

6.1.2 婴幼儿配方奶粉的配方及加工

婴幼儿配方奶粉的主要成分是牛乳或其产品粉（全脂粉或脱脂粉）、乳糖、乳清蛋白质和植物油，其形态分为粉状和液态两种。婴幼儿配方奶粉中使用的乳清蛋白质通常包含去矿物质化的乳清粉和WPC。由于人乳的灰分含量低，用于婴幼儿配方奶粉生产的所有原料（包括乳清蛋白质）对灰分含量都有严格限定。自二十世纪八十年代以来，水解乳清蛋白质（充分水解和部分水解）就已经被用于婴幼儿配方奶粉的生产当中，目的是提高消化率、减少变应原（Exl，2001）。其中，粉状配方产品比液态产品更便于运输（尤其是面对全球化的婴幼儿配方市场），价格也更便宜。表6.2是婴幼儿配方奶粉湿法配方的一个例子。

表6.2 脱脂牛乳和WPC34婴幼儿配方奶粉的湿法配方

配料	占比（%）
脱脂牛乳	57.30
乳糖	37.00
混合油	27.00
WPC34	9.40
维生素/矿物质预混料	0.50

注：资料由佛蒙特大学功能食品实验室提供。

将所有干性成分（乳糖、油、WPC34和维生素/矿物质预混料）加入脱脂牛乳中，经过均质、巴氏杀菌、蒸发、喷雾干燥等一系列加工步骤，制成婴幼儿配方奶粉。为避免加热引起某些生物活性成分损失,可将热敏性成分在喷雾干燥后进行干燥混合。乳糖作为主要添加成分,部分制造商可能会选择先干燥混合部分乳糖,来节省蒸发器和干燥机的工作负荷。液态婴儿配方奶粉是一种与奶粉具有相同成分和营养、可即时食用的产品。因此,液态配方奶被广泛应用于医院的新生儿,但它在价格上比固体奶粉更贵。液态婴幼儿配方奶也经常在家庭中使用,但没有奶粉用量大。

6.1.3 新一代婴幼儿配方奶粉的乳清蛋白质

与一般的乳清蛋白质/酪蛋白比值相比,血浆氨基酸分布更能精确地衡量婴幼儿配方奶粉与母乳的匹配程度(Picone et al., 1989)。目前,婴幼儿配方奶粉有关蛋白质的研究主要集中在乳清蛋白质上,然而母乳和牛乳的酪蛋白组成也有较大不同(表6.1)。例如,几乎一半的牛乳清蛋白质是β-LG,但母乳不含β-LG。α-LA是母乳中最丰富的蛋白质,但在牛乳中却少得多。为了使配方奶粉与母乳中的血浆氨基酸分布更加接近,就必须降低或消除婴幼儿配方奶粉中的β-LG,同时增加α-LA含量(Heine et al., 1991)。为了得到与母乳氨基酸分布更相似的配方奶粉,人们对富含α-LA或含极少量β-LG的乳清蛋白质研究越来越感兴趣。另外,这种乳清蛋白质还能减少β-LG引起的过敏风险。婴幼儿服用富含α-LA的婴幼儿配方奶后,血浆中会含有高浓度的色氨酸,这对婴儿的神经发育是很有好处的(Heine et al., 1996; Lien, 2003)。

乳铁蛋白是一种铁结合糖蛋白,是母乳中的第二种主要乳清蛋白质,但在牛乳中的含量却是微量的。与常规配方食品相比,添加乳铁蛋白的婴幼儿配方食品具有抗菌、抑制自由基、抗氧化、促进生长、改善免疫和保护双歧杆菌等特性(Schulz-Lell et al., 1991; Roberts et al., 1992; Lonnerdal and Iyer, 1995; Satué-Gracia et al., 2000; Raghuveer et al., 2002)。

甜乳清和干酪乳清制品是两种最丰富的、可用于商业化生产婴幼儿配方食品的乳清。在干酪的制作过程中,酶或化学残留物可能会出现在乳清当中,这就可能给乳清带来一些风味或营养上的问题。甜乳清中的GMP就是其中之一。GMP是凝乳酶水解κ-酪蛋白产生的片段,约占乳清蛋白质总量的20%,它可以改变婴幼儿配方奶粉中的氨基酸分布。第2章已经讨论了从乳清蛋白质中去除GMP的方法。然而,这项技术实用性差,因为去除20%的蛋白质会极大地影响乳清蛋白质的产量,而如何找到GMP的应用也将是一个问题。还有一种方法可以生产不含GMP的乳清蛋白质,即在干酪加工前将乳清分离(Rizvi and Brandsma, 2003; Nelson and Barbano, 2005; Evans et al., 2010)。微滤只能保留酪蛋白微粒和脂肪球,它们是制作干酪的成分。而乳清部分(乳清、乳糖、矿物质)可以通过膜,这个过程有利于干酪和乳清加工。对于干酪加工,微滤能够浓缩牛乳，同时大大减小后续发酵时所用发酵罐的尺寸。在乳清加工中，微滤从牛乳中生产出天然的纯净乳清。通过这种方式生产的乳清蛋白质，也被称为清蛋白（Evans et al.，2010）。

β-酪蛋白是人乳中的主要酪蛋白，因而富含β-酪蛋白的乳清蛋白质或强化乳清蛋白质越来越受到人们的关注。牛酪蛋白通常以胶束的形式存在。牛乳在4℃保存过夜后，β-酪蛋白从胶束结构解离出来转入到乳清相中（Downey and Murphy，1970），最终可以达到乳清酪蛋白质量的60%左右（Famelart and Surel，1994）。相关学者研发了一种利用冷微滤液酪蛋白酸钠提取高纯度β-酪蛋白的方法（Famelart and Surel，1994）。纯化后的β-酪蛋白可用于强化乳清蛋白质，制备婴幼儿配方奶粉。

生产婴幼儿配方食品的最终目标是能与母乳成分相匹配。随着乳清蛋白质及其分离技术与科学知识的不断进步，越来越多的乳清成分应用于生产与母乳成分相似的婴幼儿配方奶粉。

6.2 运动营养品

6.2.1 蛋白质代谢

在肌肉代谢过程中，碳水化合物、脂肪和蛋白质都起着重要作用。碳水化合物被分解为葡萄糖，然后合成糖原储存在肝脏和肌肉中，而肌糖原是肌肉在活动中容易获得的能量来源（Jeukendrup and Gleeson，2010a，b，c）。来自脂肪组织和肌肉三酰甘油的脂肪分解成脂肪酸，同时运输到线粒体中被氧化。蛋白质代谢如图6.2所示（Jeukendrup and Gleeson，2010a，b，c）。在运动过程中，肌肉蛋白质分解成游离氨基酸，然后在线粒体中氧化产生能量。氨基酸池中游离氨基酸来自日常饮食吸收和肌肉降解。未使用的游离氨基酸还可以重新合成肌肉蛋白。运

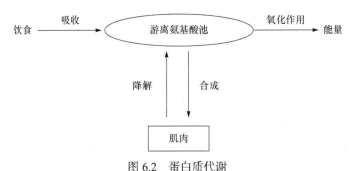

图6.2 蛋白质代谢

游离氨基酸池提供的能量源于饮食和肌肉蛋白质分解，
资料改编自 Jeukendrup and Gleeson（2010a，b，c）

动过程中，蛋白质的分解率高于合成率。运动后，只有及时摄入蛋白质和氨基酸取代氨基酸池，蛋白质合成率才能超过降解率，这对于恢复肌肉质量以及防止肌肉损伤来说，是非常重要的过程（Borsheim et al.，2002；Tipton et al.，2004，2007；Tang et al.，2007）。运动前摄入蛋白质，一般来说可能不会改善或提高运动性能和表现（Jeukendrup and Gleeson，2010a，b，c），但会促进运动后蛋白质的恢复与合成，因为它可能增加氨基酸池中的游离氨基酸含量（Phillips et al.，1997；Rasmussen et al.，2000；Wang et al.，2017）。

支链氨基酸在蛋白质代谢过程中的角色非常独特，支链氨基酸包括亮氨酸、缬氨酸和异亮氨酸。支链氨基酸被认为是一种能量来源，因为它被氧化后可以提供能量（Abumrad et al.，1989）。关于支链氨基酸的说法很多，例如，多余糖原可以减少蛋白质的分解，降低肌肉酸味，但有限的科学依据证实，只有口服支链氨基酸才具有这些优势（Jeukendrup and Gleeson，2010a，b，c）。运动后摄入支链氨基酸确实能够促进肌肉的合成（Anthony et al.，2001；Shimomura et al.，2004；Blomstrand et al.，2006）。亮氨酸不仅可以充当肌肉合成过程中所需的氨基酸构件，同时也能作为启动调节翻译起始转导途径的关键信号（Anthony et al.，2001）。食用富含亮氨酸的必需氨基酸和碳水化合物的饮食，能增强运动后蛋白质合成的抗性（Dreyer et al.，2008）。必需氨基酸的总量对肌肉恢复与合成也有重要作用，因为它们是肌肉组织的基础（Rasmussen et al.，2000）。

乳清蛋白质是必需氨基酸和支链氨基酸（尤其是亮氨酸含量高）的极佳来源（请参阅第 5 章）。蛋白质摄入的时间与营养质量同等重要（Jeukendrup and Gleeson，2010a，b，c）。乳清蛋白质被认为是一种快速消耗蛋白质。与酪蛋白相比，乳清蛋白质在胃中是可溶的，因此可以尽快通过胃到达小肠，在小肠中蛋白质被消化吸收。目前，乳清蛋白质是运动营养行业中最重要的蛋白质。

6.2.2 罐装乳清蛋白质

乳清蛋白质粉有时也被称为"罐装乳清蛋白质"或"即食型"混合粉末，是运动营养中最常见的蛋白质补充剂，因为其易于储藏、携带，同时方便加水溶解后饮用。为了在成品中获得高纯度的蛋白质，尽量最大程度减少碳水化合物（乳糖）的引入，WPC80、WPI 或它们的混合物可作为首选应用产品。表 6.3 是典型的乳清蛋白质粉饮料配方。在干燥乳清蛋白质的最后步骤中，通过喷洒卵磷脂对乳清蛋白质进行预速溶（请参阅第 2 章）。制备"预混合"蛋白粉的过程就是简单地将所有成分干燥混合，然后进行包装。为了尽可能多地提供蛋白质，典型的配方通常使用人工甜味剂代替糖来增加甜味。

表 6.3 商业化乳清蛋白质粉饮料的典型配方

配料	占比/%
乳清蛋白质粉	94~95
天然/人工香料	2~5
人造甜味剂	0.5~1
维生素/矿物质预混料	可选

注：资料由佛蒙特大学功能食品实验室提供。

6.2.3 酸化乳清蛋白质营养饮料

运动中的人一般偏爱清淡的甜味和微酸口味。牛乳的酸味还能刺激唾液分泌，同时有效地解渴。与酪蛋白不同，乳清蛋白质是一种酸性稳定的可溶性蛋白，非常适合在此类产品中应用。表 6.4 是酸性乳清粉的示例。

表 6.4 低 pH 乳清蛋白质饮料

配料	占比/%
即食分离乳清蛋白质	80.0
糖	13.4
柠檬酸	3.0
三氯蔗糖	0.05
安赛蜜	0.05
天然橙味剂	0.49
食用色素	0.01

注：资料由佛蒙特大学功能食品实验室提供。

在中性 pH 下，乳清蛋白质溶液表现为热敏性，经过巴氏杀菌/灭菌后可能无法在饮料中保持稳定。然而，当乳清蛋白质溶液通过加入有机酸或无机酸（如柠檬酸或磷酸），使 pH 低于等电点（4.6~5.0）至 3.5 时，乳清蛋白质可以在加热灭菌过程中保持稳定而不产生凝胶和沉淀。如果乳清蛋白质纯度足够高，如离子交换的 WPI，所得的蛋白质溶液将会是透明的。在添加香料等调味剂和甜味剂后，酸化的乳清饮料尝起来就像果汁一样。其中一个成功的商业化应用示例是 ISOPURE 零碳水化合物蛋白质饮料，该饮料在 20 fl oz[1 fl oz(US)=2.95735×10^{-2} dm^3] 份量的饮料中含有 40 g 纯乳清蛋白质。表 6.5 是一个乳清强化柠檬汁配方的示例。

表 6.5 液态酸性乳清蛋白质饮料配方

配料	占比/%
水	64.5
离子交换乳清蛋白质粉	5
葡萄糖颗粒	10
柠檬汁	20
水果味香精	0.50

注：资料由佛蒙特大学功能食品实验室提供。

步骤如下：
（1）将 WPI 和葡萄糖干燥混合，放入冷水中溶解；
（2）加入柠檬汁、水果味香精、色素混合均匀；
（3）巴氏灭菌；
（4）冷却并冷藏。

6.2.4 蛋白质能量补充产品

蛋白棒是一种可以提供高含量蛋白质的重要载体。与粉状或液态蛋白饮料相比，蛋白棒不需特殊工艺且易于携带。蛋白棒在远足、骑自行车、健康零食和代餐等中成为一种非常受欢迎的能量/蛋白质补充食品。如表 6.6 所示，除了蛋白质外，蛋白棒通常还含有碳水化合物、油脂、坚果或涂层物质。乳清蛋白质是蛋白棒食品配方中最常见的蛋白质来源之一，具有耐嚼性。蛋白棒的制备通常包括面团制作、切成薄片和涂层/浸渍（如果适用）。

表 6.6 蛋白棒配方

配料	占比/%
浓缩乳清蛋白质	25
固态玉米糖浆	24
蔗糖	15.00
脱脂乳粉	10
山梨糖醇	9.25
天然椰子油	7.00
椰子	3.00
卵磷脂	0.5
天然香料	0.2

注：资料由佛蒙特大学功能食品实验室提供。

步骤如下：

（1）将天然椰子油微波加热至50℃，加入蔗糖、固态玉米糖浆和卵磷脂，混合拌匀5 min；

（2）加入其他材料，混合拌匀5 min；

（3）在密封袋中存装放置过夜；

（4）将棒条切成所需的形状和大小；

（5）包装和密封。

6.3 老年人蛋白质补充剂

老年人的新陈代谢与年轻人不同，主要包括：①能量需求下降；②分解代谢大于合成代谢；③脂肪代谢减少（Breen and Phillips，2011）。因此，老年人的理想饮食需要高蛋白、低碳水化合物、低脂肪的搭配。老年人对含有氨基酸和碳水化合物混合膳食的代谢反应减弱（Volpi et al.，2000；Rasmussen et al.，2006）。近年来有人提出，老年人蛋白质摄入量的膳食营养素推荐供给量[RDA，0.8 g/(kg·d)]可能不足，在1.5 g/(kg·d)]才更合理，才能达到最佳的健康功效（Wolfe et al.，2008）。此外，由于食欲降低，老年人蛋白质的摄入量会随着年龄增长而下降。对他们而言，营养丰富的蛋白质补充剂是必需的。乳清蛋白质富含必需氨基酸（EAA）和支链氨基酸（BCAA）（尤其是亮氨酸含量特别高）。相关数据显示，与服用乳清蛋白质补充剂相比，老年人服用含有必需氨基酸的乳清蛋白质更利于肌肉蛋白的生成和吸收（Katsanos et al.，2008）。亮氨酸在蛋白质合成中起着重要的作用，然而，它必须在达到一定浓度后才能刺激或影响老年人肌肉蛋白的合成（Katsanos et al.，2008）。表6.7揭示了富含亮氨酸的蛋白质补充剂配方，建议食用量为32 g（每8 oz[①]水中），每天服用三次。

表 6.7 老年人亮氨酸强化蛋白补充剂配方

配料	占比/%
即食分离乳清蛋白质	76.12
亮氨酸	10.00
糖	10.00
天然香料	2.5
维生素/矿物质混合物	1.3
三氯蔗糖	0.08

注：资料由佛蒙特大学功能食品实验室提供。

① 1 oz = 28.349523 g。

6.4 代餐食品

代餐是旨在为那些需要代替营养和能量的人提供所有必需营养素的便捷产品。乳清蛋白质因其较高品质，很适合在这类代餐食品中应用。表6.8列出了可用作膳食替代品的蛋白质/营养饮料配方。代餐食品所处环境相对稳定，在打开包装或容器前不需要置于冰箱储存。每8 oz代餐可提供约250 kcal[①]能量、15 g蛋白质、菊粉纤维和所有必需维生素及矿物质。富含蛋白质的膳食同样有助于控制体重（Paddon-Jones et al., 2008）。

表6.8 全乳清蛋白质营养奶配方

配料	占比/%
水	73.23
玉米糖浆粉 DE 10	7.0
浓缩乳蛋白 80	5.0
植物油	4.0
浓缩乳清蛋白质 80	3.0
糖	3.0
菊粉	2.0
维生素/矿物质混合物	1.2
可可粉	1.0
天然巧克力香精	0.25
纤维素胶	0.2
磷酸二氢钾	0.1
卡拉胶	0.02

注：资料由佛蒙特大学功能食品实验室提供。

步骤如下：
（1）在70℃水中干混磷酸二氢钾、纤维素胶、卡拉胶和糖；
（2）将浓缩乳蛋白80、WPC80和玉米糖浆粉加入溶液中，充分混合拌匀；
（3）加入所有其他材料混合搅拌，直至无可见结块；
（4）将溶液在中试规模的超高温（UHT）装置中运行，最终加热温度为135℃，

---

① 1 cal（平均卡）= 4.19 J。

持续 3~6 s，并在 2000/500 psi[①]的均质条件下进行均质；

（5）将物料装入无菌瓶中。

6.5 高蛋白共生酸奶

乳清蛋白质作为一种增稠剂广泛应用于酸奶中，可以改善酸奶的质地，减少酸奶脱水（Modler and Kalab，1983；Keogh and O'kennedy，1998；Sodini et al.，2005；Li and Guo，2006）。在这些应用中，因为添加量通常很低，乳清蛋白质比蛋白质成分更能起到增稠作用。目前，由于希腊酸奶的热销，高蛋白酸奶在市场上获得人们更多的关注。通过加入 WPC 可以制成高蛋白酸奶（酸奶中总蛋白质超过 5%，普通酸奶 3%）（表 6.9）。配方和步骤如下，添加 3% WPC80 的酸奶质地坚实，口感更柔滑。还可以添加益生元和益生菌，增加其营养功能。

表 6.9 乳清蛋白质强化共生酸奶配方

配料	占比/%
全脂牛乳	87.95
浓缩乳清蛋白质 80	3.0
糖	7.0
菊粉	1.0
香精	0.5
发酵剂	0.05
益生菌	0.05

注：资料由佛蒙特大学功能食品实验室提供。

步骤如下：

（1）在 50℃条件下，向全脂牛乳中添加 WPC80、糖、菊粉和香精，将其混合直至完全溶解；

（2）85℃条件下，对牛乳进行巴氏杀菌 15 min；

（3）在水浴中冷却至 42℃；

（4）接种酸奶发酵剂和益生菌培养物；

（5）在 42℃培养 4~6 h；

（6）直至酸碱度达到 4.2~4.6，将酸奶冷却。

除了牛乳酸奶，作为蛋白质强化剂的乳清蛋白质也可用于制作其他乳酸奶，

---

① 1 psi = 6.89476 × 10^3 Pa。

如山羊酸奶（Li and Guo，2006），或使用乳清蛋白质来强化酸奶，如燕麦基共生产品。酸奶是益生元和益生菌的绝佳载体。研究表明，某些益生菌（双歧杆菌 S9、嗜酸乳杆菌 O16 和嗜酸乳杆菌 L-1）在储藏过程中生长明显（McComas and Gilliland，2003）。

6.6 总　　结

作为一种优质蛋白来源，乳清蛋白质已被广泛应用于不同食品领域，如代餐和高品质酸奶。婴幼儿配方奶粉中，乳清蛋白质也常被用来调节乳清蛋白质与酪蛋白的比例。乳清蛋白质支链氨基酸和必需氨基酸含量较高，因此在促进合成和防止肌肉组织大量流失方面具有独特的功能。而这些特性也使乳清蛋白质成为运动员、老年人以及任何想增肌或防止肌肉大幅减少人群理想的蛋白质补剂。

参 考 文 献

Abumrad, N.N., Williams, P., Frexes-Steed, M. et al. (1989). Inter-organ metabolism of amino acids in vivo. *Diabetes/Metabolism Research and Reviews* **5** (3): 213-226.

de Aguilar-Nascimento, J.E., Silveira, B.R.P., and Dock-Nascimento, D.B. (2011). Early enteral nutrition with whey protein or casein in elderly patients with acute ischemic stroke: a double-blind randomized trial. *Nutrition* **27** (4): 440-444.

Anthony, J.C., Anthony, T.G., Kimball, S.R. et al. (2001). Signaling pathways involved in translational control of protein synthesis in skeletal muscle by leucine. *The Journal of Nutrition* **131** (3): 856S-860S.

Blomstrand, E., Eliasson, J., Karlsson, H.K.R. et al. (2006). Branched-chain amino acids activate key enzymes in protein synthesis after physical exercise. *The Journal of Nutrition* **136** (1): 269S-273S.

Borsheim, E., Tipton, K.D., Wolf, S.E. et al. (2002). Essential amino acids and muscle protein recovery from resistance exercise. *American Journal of Physiology-Endocrinology and Metabolism* **283** (4): E648-E657.

Breen, L. and Phillips, S.M. (2011). Skeletal muscle protein metabolism in the elderly: interventions to counteract the 'anabolic resistance' of ageing. *Nutrition & Metabolism* **8** (1): 68.

Downey, W. and Murphy, R. (1970). The temperature-dependent dissociation of β-casein from bovine casein micelles and complexes. *Journal of Dairy Research* **37** (3): 361-372.

Dreyer, H.C., Drummond, M.J., Pennings, B. et al. (2008). Leucine-enriched essential amino acid and carbohydrate ingestion following resistance exercise enhances mTOR signaling and protein synthesis in human muscle. *American Journal of Physiology-Endocrinology and Metabolism* **294** (2): E392-E400.

European Childhood Obesity Trial Study Group (2009). Lower protein in infant formula is associated with lower weight up to age 2 y: a randomized clinical trial. *The American Journal of Clinical Nutrition* **89** (6): 1836-1845.

Evans, J., Zulewska, J., Newbold, M. et al. (2009). Comparison of composition, sensory, and volatile components of thirty-four percent whey protein and milk serum protein concentrates. *Journal of Dairy Science* **92** (10): 4773-4791.

Evans, J., Zulewska, J., Newbold, M. et al. (2010). Comparison of composition and sensory properties of 80% whey protein and milk serum protein concentrates. *Journal of Dairy Science* **93** (5): 1824-1843.

Exl, B.M. (2001). A review of recent developments in the use of moderately hydrolyzed whey formulae in infant nutrition. *Nutrition Research* **21** (1): 355-379.

Famelart, M. and Surel, O. (1994). Caseinate at low temperatures: calcium use in β-casein extraction by microfiltration. *Journal of Food Science* **59** (3): 548-553.

Guo, M. (2014). Chemical composition of human milk. In: *Human Milk Biochemistry and Infant Formula Manufacturing Technology* (ed. M. Guo), 19-32. Cambridge: Woodhead.

Heine, W.E., Klein, P.D., and Reeds, P.J. (1991). The importance of α-lactalbumin in infant nutrition. *The Journal of Nutrition* **121** (3): 277-283.

Heine, W., Radke, M., Wutzke, K.D. et al. (1996). α-Lactalbumin enriched low protein infant formulas: a comparison to breast milk feeding. *Acta Paediatrica* **85** (9): 1024-1028.

Hendricks, G.M. and Guo, M.R. (2014). Bioactive components in human milk. In: *Human Milk Biochemistry and Infant Formula Manufacturing Technology* (ed. M. Guo), 33-51. Cambridge: Woodhead.

Hoffman, J.R. and Falvo, M.J. (2004). Protein-which is best? *Journal of Sports Science & Medicine* **3** (3): 118.

Jeukendrup, A. and Gleeson, M. (2010a). Carbohydrate. In: *Sport Nutrition: An Introduction to Energy Production and Performance*, 2e (ed. A. Jeukendrup and M. Gleeson), 121-148. Leeds: Human Kinetics.

Jeukendrup, A. and Gleeson, M. (2010b). Fat. In: *Sport Nutrition: An Introduction to Energy Production and Performance*, 2e (ed. A. Jeukendrup and M. Gleeson), 149-168. Leeds: Human Kinetics.

Jeukendrup, A. and Gleeson, M. (2010c). Protein. In: *Sport Nutrition: An Introduction to Energy Production and Performance*, 2e (ed. A. Jeukendrup and M. Gleeson), 169-194. Leeds: Human Kinetics.

Katsanos, C.S., Chinkes, D.L., Paddon-Jones, D. et al. (2008). Whey protein ingestion in elderly persons results in greater muscle protein accrual than ingestion of its constituent essential amino acid content. *Nutrition Research* **28** (10): 651-658.

Keogh, M. and O'kennedy, B. (1998). Rheology of stirred yogurt as affected by added milk fat, protein and hydrocolloids. *Journal of Food Science* **63** (1): 108-112.

Kunz, C. and Lönnerdal, B. (1992). Re-evaluation of the whey protein/casein ratio of human milk. *Acta Paediatrica* **81** (2): 107-112.

Lawrence, R.A. (2000). Breastfeeding: benefits, risks and alternatives. *Current Opinion in Obstetrics and Gynecology* **12** (6): 519-524.

Li, J.C. and Guo, M.R. (2006). Effects of polymerized whey proteins on consistency and water-holding properties of goat's milk yogurt. *Journal of Food Science* **71** (1): C34-C38.

Lien, E.L. (2003). Infant formulas with increased concentrations of α-lactalbumin. *The American Journal of Clinical Nutrition* **77** (6): 1555S-1558S.

Lönnerdal, B. and Iyer, S. (1995). Lactoferrin: molecular structure and biological function. *Annual Review of Nutrition* **15** (1): 93-110.

McComas, K. and Gilliland, S. (2003). Growth of probiotic and traditional yogurt cultures in milk supplemented with whey protein hydrolysate. *Journal of Food Science* **68** (6): 2090-2095.

Modler, H. and Kalab, M. (1983). Microstructure of yogurt stabilized with milk proteins. *Journal of Dairy Science* **66** (3): 430-437.

Nelson, B. and Barbano, D. (2005). A microfiltration process to maximize removal of serum proteins from skim milk before cheese making. *Journal of Dairy Science* **88** (5): 1891-1900.

Newburg, D.S. (2001). Bioactive components of human milk. In: *Bioactive Components of Human Milk* (ed. D.S. Newburg), 3-10. Boston: Springer.

Ogbuanu, C.A., Probst, J., Laditka, S.B. et al. (2009). Reasons why women do not initiate breastfeeding: A southeastern state study. *Women's Health Issues: Official Publication of the Jacobs Institute of Women's Health* **19** (4): 268-278.

Paddon-Jones, D., Westman, E., Mattes, R.D. et al. (2008). Protein, weight management, and satiety. *The American Journal of Clinical Nutrition* **87** (5): 1558S-1561S.

Phillips, S.M., Tipton, K.D., Aarsland, A. et al. (1997). Mixed muscle protein synthesis and breakdown after resistance exercise in humans. *American Journal of Physiology-Endocrinology and Metabolism* **273** (1): E99-E107.

Picone, T.A., Benson, J.D., Moro, G. et al. (1989). Growth, serum biochemistries, and amino acids of term infants fed formulas with amino acid and protein concentrations similar to human milk. *Journal of Pediatric Gastroenterology and Nutrition* **9** (3): 351-360.

Raghuveer, T.S., McGuire, E.M., Martin, S.M. et al. (2002). Lactoferrin in the preterm infants' diet attenuates iron-induced oxidation products. *Pediatric Research* **52** (6): 964-972.

Rasmussen, B.B., Tipton, K.D., Miller, S.L. et al. (2000). An oral essential amino acid-carbohydrate supplement enhances muscle protein anabolism after resistance exercise. *Journal of Applied Physiology* **88** (2): 386-392.

Rasmussen, B.B., Fujita, S., Wolfe, R.R. et al. (2006). Insulin resistance of muscle protein metabolism in aging. *The FASEB Journal* **20** (6): 768-769.

Rizvi, S.S. and Brandsma, R.L. (2003). Microfiltration of skim milk for cheese making and whey proteins, US6,623,781B2, filed 26 September 2002 and issued 23 September 2003.

Roberts, A., Chierici, R., Sawatzki, G. et al. (1992). Supplementation of an adapted formula with bovine lactoferrin: 1. Effect on the infant faecal flora. *Acta Paediatrica* **81** (2): 119-124.

Satué-Gracia, M.T., Frankel, E.N., Rangavajhyala, N. et al. (2000). Lactoferrin in infant formulas: effect on oxidation. *Journal of Agricultural and Food Chemistry* **48** (10): 4984-4990.

Schulz-Lell, G., Dorner, K., Oldigs, H.D. et al. (1991). Iron availability from an infant formula supplemented with bovine lactoferrin. *Acta Pædiatrica* **80** (2): 155-158.

Shimomura, Y., Murakami, T., Nakai, N. et al. (2004). Exercise promotes BCAA catabolism: effects of BCAA supplementation on skeletal muscle during exercise. *The Journal of Nutrition* **134** (6): 1583S-1587S.

Sodini, I., Montella, J., and Tong, P.S. (2005). Physical properties of yogurt fortified with various commercial whey protein concentrates. *Journal of the Science of Food and Agriculture* **85** (5): 853-859.

Tang, J.E., Manolakos, J.J., Kujbida, G.W. et al. (2007). Minimal whey protein with carbohydrate stimulates muscle protein synthesis following resistance exercise in trained young men. *Applied Physiology, Nutrition, and Metabolism* **32** (6): 1132-1138.

Tipton, K.D., Elliott, T.A., Cree, M.G. et al. (2004). Ingestion of casein and whey proteins result in muscle anabolism after resistance exercise. *Medicine & Science in Sports & Exercise* **36** (12): 2073-2081.

Tipton, K.D., Elliott, T.A., Cree, M.G. et al. (2007). Stimulation of net muscle protein synthesis by whey protein ingestion before and after exercise. *American Journal of Physiology-Endocrinology and Metabolism* **292** (1): E71-E76.

Volpi, E., Mittendorfer, B., Rasmussen, B.B. et al. (2000). The response of muscle protein anabolism to combined hyperaminoacidemia and glucose-induced hyperinsulinemia is impaired in the elderly. *The Journal of Clinical Endocrinology & Metabolism* **85** (12): 4481-4490.

Walsh, H., Ross, J., Hendricks, G. et al. (2010). Physico-chemical properties, probiotic survivability, microstructure, and acceptability of a yogurt-like symbiotic oats-based product using pre-polymerized whey protein as a gelation agent. *Journal of Food Science* **75** (5): M327-M337.

Wang, W., Ding, Z., Solares, G.J. et al. (2017). Co-ingestion of carbohydrate and whey protein increases fasted rates of muscle protein synthesis immediately after resistance exercise in rats. *PLoS One* **12** (3): e0173809.

Wolfe, R.R., Miller, S.L., and Miller, K.B. (2008). Optimal protein intake in the elderly. *Clinical Nutrition* **27** (5): 675-684.

第7章 乳清蛋白质功能特性及其在食品中的应用

Cuina Wang[1], Adam Killpatrick[2], Alyssa Humphrey[2], and Mingruo Guo[3]

1. Department of Food Science, College of Food Science and Engineering, Jilin University, ChangChun, People's Republic of China
2. FoodScience Corporation, Williston, USA
3. Department of Nutrition and Food Science, University of Vermont, Burlington, USA and College of Food Science, Northeast Agriculture University, Harbin, People's Republic of China

乳清蛋白质产品广泛应用于婴幼儿配方食品、营养保健食品以及动物饲料中（Fitzsimons et al., 2008）。相较于乳清蛋白质而言，聚合乳清蛋白质具有更好的功能特性，如凝胶性、成膜性等。在一些文献中，乳清蛋白质也被称为预热乳清蛋白质、加工乳清蛋白质、热变性乳清蛋白质或变性乳清蛋白质（Vardhanabhuti et al., 2001）。在食品工业中，聚合乳清蛋白质已经被广泛用作增稠剂、稳定剂、微胶囊壁材、食用膜以及食用涂层，用来改善多种食品如香肠、乳制品、甜点、烘焙产品、凉拌酱（Elofsson et al., 1997）以及饮料、能量棒和水果（Implvo et al., 2007）的组织状态和质量。

7.1 作为食品增稠剂/胶凝剂

流体的黏度是对其抵抗剪切应力或拉伸应力引起的逐渐形变能力的一种度量。增稠剂是一种在基本不改变食品其他性能的情况下增加液体黏度的物质。通常情况下，分子量越高、支链分子越多，黏度越高。天然状态下的乳清蛋白质由于较低的分子量和近似球形的结构而具有较低的黏度（Wang et al., 2012）。然而，改性后的聚合乳清蛋白质具有较高的黏度，已被广泛用作增稠剂。图7.1所示为天然状态下乳清蛋白质和加热后热变性乳清蛋白质（有或没有盐）的结构。经过热处理后，乳清蛋白质能够形成聚合乳清蛋白质，而聚合乳清蛋白质具有较大的分子量和较高的黏度。研究表明，热聚合乳清蛋白质的固有黏度远高于天然状态下乳清蛋白质（Zhang and Vardhanabhuti, 2014）。热诱导聚合乳清蛋白质可被冷却、稀释后作为食品增稠剂添加到食品中，以改善食品组织状态。

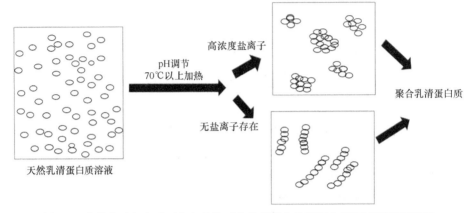

图 7.1　在盐离子存在或不存在条件下高黏度聚合乳清蛋白质溶液制备流程图
资料改编自 Bryant and Mcclements（1998）

亲水胶体的关键作用是能够凝胶化、固化流体产品。具有这种性质的胶体也被称为"胶凝剂"。乳清蛋白质的冷凝胶特性使其成为较好的凝胶剂。乳清蛋白质的冷凝胶过程分为三个步骤：首先，乳清蛋白质溶液调节至远高于乳清蛋白质等电点的 pH 处，以防止由于蛋白聚集体之间的静电排斥而导致的聚集；其次，加热乳清蛋白质以使乳清蛋白质分子变性和聚集；最后，通过改变体系的矿物质离子含量或降低 pH 至乳清蛋白质等电点后，分子之间静电斥力会降低，从而形成冷诱导凝胶（Wang et al.，2012）。在一些缺少网络结构蛋白的食品体系中，添加的聚合乳清蛋白质可发挥胶凝剂的作用，从而使其形成凝固型食品。聚合乳清蛋白质在酸性环境中能够发挥凝胶特性的能力有助于发酵食品在发酵过程中结构的形成。在食品发酵过程中，乳酸菌能够产生乳酸，降低体系 pH，从而使聚合乳清蛋白质发挥凝胶化作用（Alting et al.，2004）。发酵乳产品具有较高营养价值，在乳制品行业占有巨大市场。然而，乳清析出（也称为脱水收缩）是发酵乳制品储藏期间常见的质量缺陷。聚合乳清蛋白质添加到乳制品中，可利用乳酸菌发酵降低 pH 的作用形成酸诱导的网络结构，包埋一些小分子和水分子，从而改善发酵乳制品质地并降低乳清析出率。因此，乳清蛋白质常被用在酸奶中提高酸奶的持水力（Henriques et al.，2011）。

在 85℃、pH 7 条件下加热 10% WPI，30 min 得到聚合乳清蛋白质，其可用作牛奶、骆驼奶和山羊奶发酵乳制品的增稠/胶凝剂。Wang 等（2015）以聚合乳清蛋白质为增稠剂，成功开发了中国式老酸乳（蛋白强化酸奶）。与对照商品化样品相比，中国式老酸乳具有更低的脱水收缩性和更高的硬度及黏度。与使用琼脂或淀粉作为增稠剂的类似产品相比的另一个优势是，聚合乳清蛋白质的应用增加了最终产品中的蛋白质含量。聚合乳清蛋白质溶液（2%～8%）应用于搅拌型骆驼酸奶，可将产品黏度提高约 50%，持水量从 40%提高到 47%，脱水收缩率从 4.3%降低至 3%（Sakandar et al.，2014）。

三聚磷酸钠（STPP）是一种 FDA 允许添加到酸奶中的食品添加剂。在 pH 7～9 的条件下，蛋白质分子可与 STPP 反应，磷酸化可改善蛋白质的热稳定性和胶凝特性。磷酸化聚合乳清蛋白质可通过在 90℃条件下将乳清蛋白质溶液（10%）与 0.09% STPP 在 pH 8.4 下加热 42 min 制备而得。添加热处理的 WPI-STPP 制成的酸奶具有更优的硬度、黏性、胶着性和弹性（Cheng et al.，2017）。由于酪蛋白含量及组成的差异，发酵山羊奶很难像牛奶酸奶那样具有良好的稠度。聚合乳清蛋白质可用作胶凝剂来制备益生菌山羊奶酸奶（Wang et al.，2012）。与对照牛奶酸奶相比，添加聚合乳清蛋白质的羊奶酸奶具有相当的硬度和黏度。酸奶因带正电的钙离子中和酪蛋白微粒的负电荷，而形成簇和链组成的三维网络结构。聚合乳清蛋白质通过疏水和静电作用吸收到酪蛋白微粒中。当聚合乳清蛋白质添加到牛奶中时，随着 pH 降低，由酪蛋白、酪蛋白微粒和乳清蛋白质聚集体混合物形成凝胶（Wang et al.，2015）。通常条件下制备的聚合乳清蛋白质也可用于植物基发酵食品中。Walsh 等（2010）以聚合乳清蛋白质为增稠剂成功开发了类似酸奶的燕麦基共生产品。玉米是中国北方地区的主要谷物，但基于玉米的功能性食品非常有限，例如，由于缺乏能形成网络结构的蛋白质，难以制备基于玉米的凝固型发酵制品。然而，使用聚合乳清蛋白质（0.3%）和黄原胶（0.09%）作为协同增稠剂能够开发玉米基凝固型发酵产品（Wang et al.，2017）。

7.2 作为食品稳定剂/乳化剂

多种食品属于乳液体系，天然存在的乳液有牛乳等（Desrumaux and Marcand，2002）。乳液通常根据油相和水相彼此的相对空间分布进行分类。水包油（o/w）乳液是将油滴分散在水相中的系统。由于油和水分子之间的不利接触，乳液通常为热力学不稳定系统（Damodaran and Paraf，1997），不稳定主要包括乳脂化、絮凝/聚集和聚结等，系统最终趋于分离为两个不溶的脂肪相和水相（Damodaran，2005）。然而，乳液产品需要提高稳定性，从而延长储存期限。通过添加乳化剂克服系统的活化能可形成动力学稳定乳液。

7.2.1 乳清蛋白质乳液的表征技术

许多物理因素会影响乳液的动力学稳定性，包括液滴大小、介质黏度、分散相和连续相之间的密度差（Damodaran，2005）。降低液滴尺寸和密度差、增加体系黏度能够提高乳液稳定性。因此，这些参数常被用来表征乳清蛋白质乳液。

1. 粒径

粒径和粒径分布通常使用动态光散射方法来测定。油滴的平均直径表示为体积平均直径（d_{43}）或平均油滴直径（d_{32}）。跨度（span）用来表示粒径分布宽度。

$$d_{43} = \sum n_i d_i^4 / \sum n_i d_i^3 \quad (7.1)$$

$$d_{32} = \sum n_i d_i^3 / \sum n_i d_i^2 \quad (7.2)$$

$$\text{span} = (d_{90} - d_{10})/d_{50} \quad (7.3)$$

其中，n_i 是直径为 d_i 的粒子数；d_{10}、d_{50} 和 d_{90} 分别是体积等效粒径 10%、50% 和 90% 的直径。

2. 剪切黏度

通常条件下，以剪切速率为变量，使用流变仪测量乳液介质的剪切黏度。

3. 乳化稳定性

将一部分乳液倒入圆柱形玻璃试管中，密封并在 25℃ 下保存一段时间，通过测量试管底部乳清相高度随着时间变化得到（Kuhn and Cunha，2012）。

4. 氧化稳定性

乳清蛋白质乳液过氧化物值（PV）也是衡量乳液储存过程中稳定性的指标之一。过氧化物值一般采用 IDF 标准方法中的分光光度法测定。

7.2.2 乳清蛋白质乳液的制备

乳清蛋白质由于其分子同时含有亲水（憎油）性的极性基团和亲油（憎水）性的非极性疏水基团而属于两性物质。乳清蛋白质能在油水界面形成黏弹性膜，形成分散良好的乳状液滴。β-LG 和 α-LA 吸附在油水界面上，能形成稳定的乳液。乳清蛋白质乳液在模拟胃液条件下是稳定的，而在模拟肠液条件下则可分解（Mantovani et al.，2017）。与其他常用乳化剂相比，乳清蛋白质稳定乳液的稳定性略低于同样条件下用酪蛋白稳定的乳液（Hunt and Dalgleish，1994），与阿拉伯胶相比，较低浓度时，乳液液滴粒径较小，但对环境胁迫的稳定性较差（Charoen et al.，2011）。

1. 乳清蛋白质乳液的制备方法

油包水乳液是乳清蛋白质乳液的主要类型。图 7.2 为制备乳清蛋白质乳液的一般流程。通常情况下，将乳清蛋白质或乳清蛋白质与胶体的混合物溶解于水或 PBS 缓冲液中，搅拌至完全分散。可加入叠氮化钠作为抗菌剂，油相缓慢加入水相中。通过传统的剪切均质得到粗乳液，剪切均质的目的是将油相分散到水相中。为了制备质地优良、稳定性较好的乳液，可以采用不同循环次数的高压均质或微射流处理乳液。

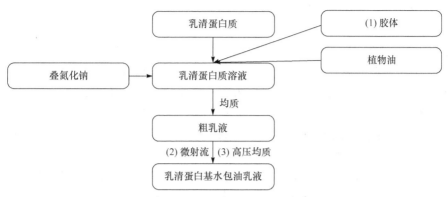

图7.2 乳清蛋白质作为稳定剂的乳液制备流程图
（1）（2）和（3）为可选步骤

均质条件包括均质压力和均质次数，对乳化液的稳定性有很大影响。将豆油分散在 WPC 乳液中喷雾干燥，得到含油量为 20%~75% 的乳粉，在均质压力 10~50 MPa 范围里，乳化油滴粒径随均质压力的增大而减小（Hogan et al.，2001）。将均质压力从 25 MPa 提高到 100 MPa，显著降低了乳液的粒径[乳液制备条件：WPC（10%）、大豆油（0%~5%）和大豆卵磷脂（0%或5%）]（Yan et al.，2017）。在亚麻油/WPI 乳液中，均质压力从 20 MPa 增加到 80 MPa，均质 3 次，乳液的平均粒径降低。在 80 MPa 压力下，均质循环次数的增加导致了高分子质量聚合物（>200 kDa）的形成，这有利于乳液黏度的增加（Kuhn and Cunha，2012）。其他研究者（Ng et al.，2017）也证实了可以通过提高均质压力和均质循环次数来减小 WPI 纳米纤维-海藻酸盐复合物稳定的棕榈油酸乳液粒径，均质压力的增加和均质循环次数的增加会导致 zeta 电位负值更大。

超高压均质仪还可用于制备优质的水包油乳液。经过超高压均质（25℃，100 MPa 或 200 MPa）处理的水包油乳液中脂肪液滴的粒径为 100~200 nm。与传统方法比，超高压均质处理乳液在 10 天存储过程中氧化速率显著降低（Hebishy et al.，2017）。Desrumaux 和 Marcand（2002）采用 350 MPa 的均质化工艺生产了以 WPC（1.5%）为乳化剂的水包油乳液（油相为 20%向日葵油），随着压力的增加，乳液的结构和质地发生了显著的变化，这可能是蛋白质构象发生改变而造成的。

2. 乳清蛋白质乳液组成

蛋白质的分子性质如疏水性、构象稳定性、电荷和摩尔质量对蛋白质作为乳化剂稳定的乳液性质具有重要的影响。通常情况下，天然乳清蛋白质稳定的乳液比热处理乳清蛋白质稳定的乳液有更好的稳定性。加热导致乳清蛋白质乳化性能显著降低（Kiokias et al.，2007）。

蛋白质含量和油量对乳液的理化性质和黏弹性有很大影响，主要通过液滴的粒径分布、乳脂化、氧化稳定性和流变性来实现。在含有不同蛋白质浓度（0.2%、1%、2%）、0.2%黄原胶以及不同油相体积分数（5%、20%、40%）的水包油乳液中，增加WPI浓度会显著影响液滴尺寸、表面电荷和氧化稳定性，但对乳脂稳定性几乎没有影响。同时，研究者发现油相体积分数在不同的蛋白质添加水平上对乳液特性具有不同的影响。在0.2% WPI条件下，增加油相体积分数会大大增加液滴尺寸，但对表面电荷没有明显影响。在WPI为1%或2%时，增加的油相体积分数对液滴尺寸的影响较小，但导致产生更多的负表面电荷（Sun and Gunasekaran，2009）。Hogan（2001）等制备了含有WPC（5%）和体积分数从20%至75%的大豆油的乳液，发现含油量不会影响乳液油滴尺寸大小，但乳液蛋白荷载量会随着油/蛋白质比例的增加而降低。

7.2.3 乳清蛋白质乳液的稳定性

1. 热处理稳定性

在实际应用中，对蛋白质稳定的乳液进行蒸煮、巴氏灭菌或灭菌等热处理具有重要的意义。乳清蛋白质是热敏感型球状蛋白混合物。β-LG、α-LA和BSA这三种主要乳清蛋白质的变性温度分别为78℃、62℃和64℃（Bryant and Mcclements，1998）。乳清蛋白质稳定乳液的稳定性会由于吸附在相同或不同液滴上的蛋白质之间的相互作用而发生改变（Dickinson，1994）。因此，乳清蛋白质乳液的热稳定性低于酪蛋白乳液的热稳定性。乳清蛋白质乳液对热的敏感性与介质pH、离子强度、加热温度和加热时间高度相关。当盐浓度为100 mmol/L，在30~90℃加热时，用2% WPI稳定的水包玉米油乳液在WPI的等电点（pH 5）附近具有糊状结构。但是，乳液在pH 6或4时仍保持流体状。pH 7的乳液（特别是在高盐浓度下）在加热至70~80℃时会导致絮凝，但在pH 3时的影响较小（Demetriades et al.，2010）。加热的乳清蛋白质稳定乳液在pH 7.0、75℃时会导致絮凝（Sliwinski et al.，2003）。

2. 离子强度的影响

液滴表面结合相反离子或静电屏蔽作用可能导致液滴之间的静电排斥力降低，从而进一步导致带电荷蛋白质离子聚集。在高盐离子浓度条件下，由于液滴之间静电和疏水相互作用的变化，WPI乳液将发生大量液滴聚集现象（Charoen et al.，2011）。在无盐离子的情况下，由于较低的电荷作用，乳清蛋白质稳定的乳液液滴在等电点附近（4.5<pH<5.5）发生聚集，导致出现乳脂不稳定现象。在较高的pH条件下，当向乳液中添加$CaCl_2$时，将会发生液滴聚集现象（Kulmyrzaev et al.，2000）。KCl浓度超过10 mmol/L时，加快了乳清蛋白质稳定乳液的聚集趋

势（Kulmyrzaev and Schubert，2004）。钙添加的效果取决于乳清蛋白质浓度，对于 0.5% WPC 乳液，乳化前添加高于 3 mmol/L 的 $CaCl_2$ 会导致液滴直径增大和稳定性降低。对于 3% WPC 乳液，在 $CaCl_2$＞15 mmol/L 时，乳液稳定性显著下降（Ye and Singh，2000）。

3. 降低油的氧化敏感性

在加工过程中，一些不饱和脂肪酸由于高度不饱和而易于氧化（Kuhn and Cunha，2012）。据报道，乳清蛋白质具有抗氧化活性，这可能对于含有不稳定的容易氧化的成分作为分散相的体系非常有益。天然乳清蛋白质比酪蛋白酸钠和吐温 20 乳化剂能更有效抑制水包油型乳剂的脂质氧化（Kiokias et al.，2007）。与天然乳清蛋白质相比，热处理乳清蛋白质（尤其是高于 60℃）具有较低的氧化敏感性（Kiokias et al.，2007）。

7.2.4 乳清蛋白质/水胶体乳液稳定性

蛋白质和多糖是调控食品中乳液质构以及稳定性的最重要的两种生物大分子物质。乳清蛋白质被用作乳化剂。由于增稠作用，一些多糖作为稳定剂添加到乳液中，从而增加了连续相的黏度，并通过阻止液滴的移动而增强乳液的稳定性（Sun et al.，2007）。因此，可将乳清蛋白质和多糖的特性结合，将二者复合使用（Dickinson and Euston，1991）。通常情况下，乳清蛋白质和多糖可以通过共价结合，或通过非共价键结合，如静电相互作用。

1. 多糖与乳清蛋白质的共价复合物

果胶是一种阴离子多糖，已被用于改善乳清蛋白质稳定乳液的特性。果胶对乳清蛋白质乳液的影响取决于介质 pH 和果胶类型，在 pH 5.5 条件下与低甲氧基果胶共价结合，可通过增强的静电以及空间位阻作用来提高乳清蛋白质乳液稳定性，在 pH 4 条件下，乳化性能下降（Neirynck et al.，2004）。WPI-甜菜果胶共价复合物乳液具有更好的物理稳定性，如减小的液滴尺寸和更均匀的液滴尺寸分布（Xu et al.，2012）。WPI-甜菜果胶共价复合物还改善了乳清蛋白质乳液的热稳定性，0.5% WPI-低甲氧基果胶共价复合物乳液（10%油）在 80℃和 120℃加热长达 20 min 时显示出热稳定性显著改善（Setiowati et al.，2017）。通过干热法将乳清蛋白质与高甲氧基果胶结合，可明显降低乳清蛋白质的溶解度，进而降低乳清蛋白质的乳化性能（Neirynck et al.，2004）。

2. 多糖与乳清蛋白质非共价复合物

在溶液中，乳清蛋白质和阴离子多糖通常通过静电相互作用来提高乳液液滴的稳定性。在蛋白质等电点以下和以上，果胶对蛋白质稳定化乳剂均具有提高乳化稳定的作用。在 pH 4.0 条件下，较高浓度的果胶吸附到乳清蛋白质上会引起电荷逆转，从而产生较小的液滴尺寸。在 pH 5.5 条件下，果胶积聚在液滴界面上，当添加果胶后，蛋白表面电荷逐渐变为负值，从而导致液滴尺寸更小，乳液稳定性显著提高（Neirynck et al.，2007）。

黄原胶（XG）是一种良好的稳定剂，已被用于通过增加乳液黏度来稳定乳液。在 WPI 稳定鲱鱼 o/w 乳液中，0.5%黄原胶体系的屈服应力为 1.54 Pa，稳定性显著提高（Sun et al.，2007）。

在溶液状态下，WPI 和阿拉伯胶（GA）混合物可通过电荷-电荷相互作用提高乳液稳定性。阿拉伯胶的添加可增加乳清蛋白质乳液的表观黏度（Ibanoglu，2002）。pH 7 条件下，WPI 和阿拉伯胶的芥花油乳液比单独的 WPI 或阿拉伯胶乳液稳定性更好。乳液中液滴尺寸小于 1 μm，并在 25℃条件下储藏 1 个月无明显增长（Klein et al.，2010）。将较低质量比的阿拉伯胶/壳聚糖复合物掺入乳清蛋白质乳液中导致相分离。提高多糖的质量比可产生高黏度的液滴网络，从而防止甚至抑制相分离。进一步增加阿拉伯胶/壳聚糖质量比，乳液液滴被固定在不溶性三元基质形成的簇中（Moschakis et al.，2010）。

7.2.5 其他乳化剂对乳清蛋白质乳液稳定性的影响

食品工业中，许多表面活性剂和蛋白质或者多种蛋白质会共同使用。大豆卵磷脂是一种良好的乳化剂，具有两亲分子结构，分子中既包含脂肪酸基团形式的亲脂部分，又包含磷酸酯基形式的亲水基团。大豆卵磷脂与乳清蛋白质的共存和相互作用会影响乳清蛋白质的乳化性。卵磷脂的添加量在 0.25%～2%之间会降低 WPI 稳定的牡丹籽油乳液的液滴尺寸和界面黏弹性。卵磷脂和乳清蛋白质在界面上相互作用，使吸附更紧密，提高了乳液的稳定性（Wang et al.，2017）。在乳化过程中，当乳清蛋白质和酪蛋白一起使用时，在界面处会有一些竞争性蛋白质吸附。用 WPC 和乳清蛋白质：酪蛋白的混合物（1∶1）分别作为乳化剂制备水包油乳液（30%豆油），两种乳液具有相似的平均油滴直径（d_{32}）和总表面蛋白浓度。在两种蛋白质的总浓度较高时，酪蛋白的存在会降低乳液的稳定性（Ye，2008）。

7.3 脂肪或乳品替代品

脂肪能够影响食品加工过程，影响食品的口感和质构特性。因此，在降低食品中脂肪含量时应考虑降低脂肪含量对食品特性的影响。如果降低脂肪含量，则应在食品中添加脂肪替代品。WPC 已成为一种非常有效的脂肪替代品，乳清蛋白质的能量为 4 kcal/g，远低于脂肪的能量（9 kcal/g）。与碳水化合物和脂肪相比，蛋白质在满足饱腹感方面更有效，从而有可能减少食物的总摄入量（Chung et al.，2014）。

蛋白质微粒可通过微粒化过程制备得到，这些蛋白质微粒具有能够提供类似脂肪的口感所必需的大小和形状（O'Brien et al., 2003）。乳清蛋白质可以模仿人对真实脂肪的口感和质感的原因之一是聚合乳清蛋白质可产生与脂肪小球相似粒径的微凝胶。基于乳清蛋白质的脂肪替代品现已上市，并因其生产优质低脂产品的潜力而得到积极推广。Simplesse®和 Dairy-Lo®是商业化 WPC 微粒的例子。Simplesse®是一种基于蛋白质（乳清蛋白质）的脂肪替代品，最初于 1988 年以干粉形式投放市场。乳清蛋白质微粒分散体是通过加热乳清蛋白质使其部分聚集而形成的，从而产生与脂肪相似的乳脂状和浓稠度。Simplesse®可作为蛋白质消化，从热量角度来看，它只产生 1.0～1.3 kcal/g 的热量，普通脂肪产生 9.0 kcal/g 的热量。

乳脂可作为脂溶性风味物质和营养素的载体，还影响酸奶的质地，如凝胶化和脱水收缩，并且对于感官特性（如光泽、颜色和味道）非常重要（Liu et al., 2016）。在 85℃、pH 8.5 的条件下加热浓度为 12.5%的 WPC 溶液（10%纯蛋白）30 min 制备的微粒化 WPC，已被用作脱脂山羊酸奶中的有效脂肪替代品（Zhang et al., 2015）。含有鱼油的 WPC 微粒可通过均质（10 MPa 或 20 MPa），在 85℃、pH 8.5 下加热 30 min 制备得到。微粒化 WPC 大部分（约 85%）分布在（1106±158）nm 范围内。将聚合的微粒化 WPC（12%，每 100 g 含 1300 mg DHA 和 EPA）加入到酸奶中，可使其具有与全脂酸奶相似的感官和质地特性（Liu et al., 2016）。脱脂甚至无脂干酪或者肉类具有卓越的感官特性并不容易。因为其蛋白质含量过高，交联反应更易发生。随着脂肪含量的减少，蛋白质基质变得更加致密，干酪的质地变得更有嚼劲（Kavas et al., 2004）。应对这些挑战的一种方法，就是在低脂猪肉香肠中应用高胶凝性的乳清蛋白质（Lyons et al., 1999）。

7.4　疏水性营养食品载体

β-LG 是脂蛋白，具有结合各种疏水性物质（如脂肪酸、血红素、玫瑰树碱、芳香烃和致癌烃）的能力。乳清蛋白质是亲脂性保健食品合适且有效的载体。

7.4.1　类胡萝卜素

类胡萝卜素是在植物和藻类以及细菌和真菌中发现的一类天然色素（Ribeiro et al., 2006）。目前已知的类胡萝卜素超过 600 种，分为叶黄素类（含氧）和胡萝卜素（纯烃，不含氧）。β-胡萝卜素和番茄红素是胡萝卜素，而虾青素、岩藻黄质和叶黄素是叶黄素类。这些疏水性功效成分是脂溶性的，主要储存在动物脂肪组织中。由于其分子结构中存在双键，类胡萝卜素很容易被氧化剂、光和热降解，从而导致产品质量下降。就其低生物利用度而言，在水果和蔬菜中较大的粒径可能是主要因素，而类胡萝卜素也以晶体形式存在或以蛋白质复合物的形式存在（Ribeiro et al., 2006）。

保护类胡萝卜素免于氧化降解的一种方法可能是将其掺入水包油乳液的油相中。除了最常用的形成乳液的方法外，疏水性功效成分也可与乳清蛋白质复合形成纳米颗粒，从而提高生物活性和/或提高稳定性。在本节中，我们主要关注上述类胡萝卜素功效成分与乳清蛋白质之间的相互作用，并阐明乳清蛋白质可用作类胡萝卜素载体的机理。

1. 胡萝卜素

β-胡萝卜素是一种橘红色色素，是维生素 A 的已知来源，具有良好的抗氧化剂和清除自由基的能力。溶解性低和稳定性差等原因，限制了β-胡萝卜素在食品工业中的应用。当β-胡萝卜素与β-LG 以 1∶2 的配体/蛋白质比例复合时，可以防止β-胡萝卜素的氧化反应。酯化和烷基化β-LG 使β-胡萝卜素/蛋白质结合比例变为 1∶1（Dufour and Haertlé, 1991）。维生素 A 是一种脂溶性化合物，对许多生物过程包括视力、胎儿生长、免疫应答、细胞分化和增殖等具有至关重要的作用。许多研究已经表明，β-LG 可与维生素 A 分子以 1∶1 的比例紧密结合（Visentini et al., 2016）。

2. 叶黄素类

虾青素是一种类胡萝卜素，分子内含有 13 个共轭双键和 2 个官能团（酮基和羟基）。虾青素是一种良好的抗氧化剂，具有清除自由基和猝灭单线态氧的能力（Ribeiro et al., 2005）。乳液体系可用来提高虾青素的生物利用率并防止其氧化。Shen 等（2018）用 WPI 和聚合乳清蛋白质（PWP）通过乳化蒸发技术制备虾青素纳米分散体系。乳清蛋白质稳定的虾青素纳米分散体能够抵抗胃蛋白酶的降解作用，但在胰蛋白酶消化后易于释放虾青素。WPI 和 PWP 稳定的纳米分散体系使 Caco-2 细胞对虾青素的表观渗透系数（P_{app}）分别提高了 10.3 倍和 16.1 倍。

岩藻黄质是一种在可食用的棕色海藻中发现的海洋类胡萝卜素，是一种有效的可预防肥胖症及其相关 2 型糖尿病的天然化合物。岩藻黄质是一种疏水性颜料，具有独特的结构，包括稀有的二烯键和 5,6-单环氧化物。纯度高的岩藻黄质极易被氧化（Kraan and Dominguez, 2013）。岩藻黄质可以与乳清蛋白质结合并形成纳米复合物。BSA/β-LG/α-LA 与岩藻黄质能够形成纳米级（<300 nm）的聚集体。结合位点数目约等于 1。三种蛋白对岩藻黄质结合能力按照递减的顺序为 BSA>β-Lg>α-La。三种蛋白与岩藻黄质的结合都属于自发过程，驱动力主要包括范德瓦耳斯力、氢键和疏水相互作用的非共价相互作用（Zhu et al., 2017）。

7.4.2 多酚

多酚是一类天然的、合成的或半合成的有机化学物质，分子中主要含有大量的酚结构单元。多酚通常具有强大的抗氧化性能，但稳定性很低。在本节中，我们主要讨论乳清蛋白质在保护多酚免受不利环境影响方面的作用，以姜黄素和花青素作为多酚类物质模型。值得注意的是，多酚类物质通常通过疏水力与蛋白质相互作用，然后通过氢键稳定。

姜黄素水溶性低、光稳定性差。在中性和碱性条件下，姜黄素的共轭二烯结构易降解。β-LG/海藻酸钠静电纳米复合物可用来封装和输送姜黄素。姜黄素的胶囊化显著提高了其分散性及热稳定性。与空白样品相比，姜黄素纳米复合物储存时间明显延长。载有姜黄素的纳米复合物可抵抗胃液降解，并在模拟的肠液中持续释放（12 h 释放率为 78.5%）（Mirpoor et al., 2017）。

花青素（ACN）是水溶性的天然食用色素，存在于各种水果、蔬菜和花卉中。它们通常用作功能性食品和饮料中的着色剂，并在营养保健品中具有潜在的治疗作用（Miguel, 2011）。花青素是一种强大的抗氧化剂。然而，花青素化学稳定性较差，半衰期较短。由于其在 pH、温度、光、氧气和酶等条件下容易降解，所以花青素的生物利用率较低（Ge et al., 2018）。与乳清蛋白质分子结合（主要为氢键和疏水相互作用），花青素的稳定性得到了显著提高。在加速储藏实验中，乳清蛋白质，尤其是变性乳清蛋白质，能显著改善模型饮料的颜色。乳清蛋白质与花青素上的黄酮离子的缔合保护了黄酮离子的 2 位免受亲核攻击（Chung et al., 2015）。为了提高花青素的稳定性，乳清蛋白质已作为不同来源花青素的微胶囊壁材，具体内容将在下节进行讨论。

7.5　微囊化壁材

微囊化是一种用于保护多种生物分子（小分子和蛋白质）以及细菌、酵母和动物来源细胞的技术（Borgogna et al., 2010）。微囊化是一个微小的颗粒或液滴被用作小的胶囊壁材包覆的过程。微囊化技术适于制备含有敏感成分的蛋白质基微胶囊，以达到控释和提高稳定性的作用（Cho et al., 2003）。研究者对乳清蛋白质和聚合乳清蛋白质作为壁材应用已进行了大量研究和开发（表 7.1）。

表 7.1　乳清蛋白质作为微胶囊壁材应用

类别	芯材
油类	丁酸乙酯、辛酸乙酯、亚麻油、乳脂、亚麻油、香草香精、咖啡油
益生菌	双歧杆菌 Bb-12、植物乳杆菌 A17、嗜酸乳杆菌 Ki、副乳杆菌 L26、短双歧杆菌 R070、长双歧杆菌 R023、长双歧杆菌 1941、嗜酸乳杆菌 LA-5、干酪乳杆菌 431、鼠李糖乳杆菌 CRL 1505
活性物质	β-胡萝卜素、人参皂苷、花青素、白藜芦醇、虾青素、茶碱、鱼油、咖啡因
维生素	维生素 A、维生素 D_3、胆盐水解酶

7.5.1 乳清蛋白质基风味化合物和脂质微胶囊的制备

1. 微胶囊化评价参数

香气和风味成分的微囊化是食品体系中微囊化最重要的应用之一。脂质的包封是一种无需抗氧化剂即可防止氧化的物理手段。在微囊化过程中，含高比例的挥发物和较高的保留率是重要指标。成功的喷雾干燥封装技术取决于在处理和储藏过程中芯材（尤其是挥发物）的高度保留以及粉末颗粒表面油（挥发物或非挥发物）含量最少。

1）微胶囊化率

微胶囊化率（MY）定义为最终干燥的微胶囊中芯材与乳液中芯材的比例。

$$MY=\frac{微囊中芯材质量(g/100g固体)}{乳液中芯材质量(g/100g固体)} \quad (7.4)$$

2）微囊化效率

微囊化效率（MEE）定义为通过石油醚/己烷提取的微囊表面油（总油与提取油之间的差）与总油的质量比。

$$MEE=(TO-EO) \times 100/TO \quad (7.5)$$

其中，TO 是总油；EO 是提取油。

3）脂质氧化

脂质氧化通常是测定完全提取脂肪中的过氧化物值。

4）水分含量

通过甲苯蒸馏来测定水分。

5）挥发油的储存稳定性

顶空气相色谱法用于测定脂质氧化产生的一些物质，可直接用于测量挥发油含量。

2. 微囊化工艺

喷雾干燥是食品工业中使用最广泛的微囊化技术，通常用于制备干燥、稳定的食品添加剂和调味剂（Desai and Park，2005）。使用乳清蛋白质作为壁材制备粉末状微胶囊通常包括乳液制备以及乳液喷雾干燥两个步骤。图 7.3 所示为食品香料和油类的喷雾干燥微囊化流程图。壁材和芯材的特性、乳液特性以及干燥工艺条件影响芯材的效率和保留率（Jafari et al.，2008）。使用乳清蛋白质与碳水化合物的混合物（含有干燥后的葡萄糖浆或物理改性的抗性淀粉的葡萄糖水合物）或加热后的这些混合物作为壁材的金枪鱼油粉（25%和 50%油相）核磁共振波谱结果显示，微囊化油粉的溶解特性取决于所用碳水化合物的类型以及蛋白质/碳水化合物基质在微囊化前是否经过热处理（Burgar et al.，2009）。用喷雾冷冻法代替喷雾干燥法制备的粉末状微胶囊能显示出更好的热稳定性（Hundre et al., 2015）。

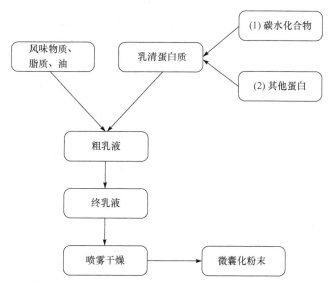

图7.3 利用喷雾干燥技术制备乳清蛋白质基风味物质、脂质和油微胶囊流程图
（1）和（2）为可选步骤，资料改编自Jafari et al. （2008）

3. 风味物质和脂质微胶囊的组成

喷雾干燥微囊化技术主要包括分散液制备、分散液均质化、进料分散液雾化和雾化颗粒脱水四个步骤。含油微胶囊的组成即为乳液的组成。表7.2列出了一些以乳清蛋白质作为壁材通过喷雾干燥技术制备风味物质和脂质微胶囊的例子。WPC、WPI或乳清蛋白质与糖的混合物通常用作壁材。WPC被认为是包埋共轭亚油酸（CLA）的有效壁材。基于乳清蛋白质的CLA微胶囊的包封效率为89.60%，表面油浓度为1.77 g/100 g样品，微胶囊在60天的储藏期内表现出良好的抗氧化稳定性（Jimenez et al., 2004）。亚麻籽油被认为是ω-3脂肪酸的可靠来源，对预防卒中和心脏病具有重要作用。亚麻籽油颗粒表面被蛋白层覆盖，微胶囊用于降低储存期间油相的氧化速率（Partanen et al., 2008）。WPC作为壁材制备四种不同油相浓度（相对于总固体为10%、20%、30%和40%的油相）的微胶囊，结果表明，增加油相浓度，降低了包封效率并加强了脂质氧化作用（Tonon et al., 2012）。

与脱脂奶粉（SMP）相比，WPC是一种更有效的包埋香菜油的壁材，包封效率为78%，而SMP则为74%。与SMP相比，基于WPC的微囊化颗粒还呈现出更少裂纹和孔眼的微观结构（Bylaite et al., 2001）。

表 7.2 以乳清蛋白质为壁材采用喷雾干燥法制备风味物质和脂质微胶囊

芯材	壁材	喷雾干燥工艺	结果	参考文献
共轭亚油酸	芯材与壁材（30% WPC 溶液）比例：1:4（w/w）	进风温度：200℃；出风温度：110℃	包埋效率：89.60%。表面油浓度：1.77 g/100 g。a_w=0.743～0.898 条件下储藏，微胶囊有良好的氧化稳定性，保藏期可达 60 天	Jimenez et al., 2004
亚麻油	10%乳清分离蛋白和质量为干燥样品 40%的油	进风温度：180℃；出风温度：90℃	在 37 ℃条件下储藏 9 周，喷雾干燥的油比散装油具有更低的氧化速率	Partanen et al., 2008
亚麻油	总固形物（油和乳清蛋白质）：30%，油相占总固形物比例：10%～40%	进风温度：180℃；出风温度：100℃	包埋效率：37%～70%。PV：1～2 meq 过氧化物/kg 油。油相浓度：10%～40%	Tonon et al., 2012
葛缕子精油	乳清蛋白质与麦芽糊精比例：1:9，30%。葛缕子精油占固形物比例：15%	进风温度：180℃；出风温度：90℃	包埋效率：82%～86%。与未包埋油相比，包埋油相产生了较少量的氧化产物	Bylaite et al., 2001
丁酸乙酯和辛酸乙酯	乳清蛋白质与乳糖比例：1:1	进风温度：160℃；出风温度：180℃	保留率高达 92%	Rosenberg and Sheu, 1996
亚麻籽油	壁材：17%阿拉伯胶、66%麦芽糖糊精（MD）和 17%的三元混合物；壁材和芯材比例：4:1	进风温度：175℃；出风温度：75℃	包埋效率：87.8%	Gallardo et al., 2013
亚麻油	总固形物质量：30%，亚麻油占总固形物的 20%	进风温度：180℃；出风温度：110℃	包埋效率：60%，麦芽糊精和乳清蛋白质作为壁材生产的微胶囊在 4 周储存期内未见氧化产物	Carneiro et al., 2013

4. 乳清蛋白质/多糖为壁材的油脂微囊化

由于单一封装剂的壁材特性并不理想，因此使用多种大分子复合物作为壁材成为一种趋势。与单独的乳清蛋白质相比，乳清蛋白质和其他多糖复合物将改善微胶囊特性。与单独乳清蛋白质相比（76%），WPI 和乳糖（1:1）的复合壁材制备的丁酸乙酯和辛酸乙酯微胶囊具有更高的保留率（92%）（Rosenberg and Sheu, 1996）。用海藻酸盐包被乳清蛋白质层可产生更稳定的乳液。但是，这种作用取决于海藻酸钠的浓度和介质 pH（Fioramonti et al., 2015）。由 WPI、阿拉伯胶和麦芽糖糊精的三元复合物制成的微胶囊对芯材抗氧化保护作用可高达 90%。研究人员同时指出三元复合物为壁材的微胶囊具有表面光滑的球形结构，并且这种结构储藏 10 个月后仍保持不变（Gallardo et al., 2013）。乳清蛋白质/麦芽糊精复合物作为壁材虽然有较低的包封效率，但能更好地保护活性物质，防止脂质氧化

(Carneiro et al., 2013)。用麦芽糖糊精部分取代 WPC 可增加喷雾干燥过程中挥发性物质的保留率，并能更好保护固化胶囊免受氧化（Bylaite et al., 2001）。γ-环糊精/浓缩乳清蛋白质/黄原胶（0.8∶0.2∶0.5）三元复合物作为壁材包埋鱼油，能产生较小的颗粒，负载率为 40%，鱼油封装效率为 80%。此外，胶囊化鱼油的气味强度降至其原始值的 30%（Na et al., 2011）。

7.5.2 益生菌微囊化

益生菌是指能够通过改善肠道微生态平衡对宿主产生有益作用的活的微生物（Kim et al., 2008）。每克产品含有 10^6～10^7 个活的益生菌才能在人体中发挥有益的作用（Mandal et al., 2006）。然而，益生菌在消化前后对环境都非常敏感。益生菌进入人体消化系统后，胃部酸性环境和十二指肠分泌的胆汁盐影响益生菌活性（De Castro-Cislaghi et al., 2012）。研究表明，影响发酵乳制品中益生菌生存能力的因素有很多，包括滴定酸度、pH、过氧化氢、溶解氧含量、储存温度、发酵剂及附属发酵剂、乳酸和乙酸的浓度，甚至包括乳清蛋白质浓度（Anal and Singh, 2007）。因此，可能需要微囊化技术提高益生菌的生存活力，并实现在胃肠道中的定向递送。

1. 用于评估乳清蛋白质基益生菌胶囊的参数

1）粒径

微胶囊的尺寸分布可利用激光粒度分析仪进行测量。

2）游离和封装益生菌计数

游离益生菌可通过平板计数法直接计算。微囊化益生菌可以通过均质等方法释放益生菌，随后通过平板计数法进行计数。

3）益生菌在模拟胃肠道条件下的存活特性

微囊化益生菌的主要目标之一是保护其免受人体胃肠道不利环境的影响。人工消化液通常使用胃蛋白酶和胰蛋白酶制备。通常情况下，对于制备人造胃液，将胃蛋白酶（3 g/L）悬浮在无菌氯化钠溶液（0.5%）中，并用浓 HCl 将悬浮液的 pH 调节至 2.0。通过将胰蛋白酶（10 g/L）和胆汁盐（3.0 g/L）悬浮在无菌氯化钠溶液（0.5%）中，并用 0.1 mol/L NaOH 调节 pH 至 8.0 来制备模拟肠液（Arslan et al., 2015）。

4）益生菌储存稳定性

将益生菌在不同的温度及湿度条件下保存特定的时间，并定期对存活的益生菌进行计数。

2. 益生菌微囊化制备技术

益生菌通常通过喷雾干燥、冷冻干燥和挤出法进行封装，如表 7.3 所示。

表 7.3 益生菌微囊化中常用技术

壁材	益生菌	微囊化技术	结果	参考文献
分离乳清蛋白质：10%~30%（w/w）	植物乳杆菌A17和植物乳杆菌B21	喷雾干燥技术，进风温度：110℃，空气流速：70 m³/h，进料速度：6.6 mL/min	植物乳杆菌A17存活率：91.7%，植物乳杆菌B21存活率：88.7%	Khem et al., 2016
液态乳清：固形物含量6.1 g/100 g	双歧杆菌Bb-12	喷雾干燥技术，进风和出风温度分别为150℃和50~60℃	4℃储藏12周：益生菌数量稳定（>10⁹ CFU/g），下降幅度低于10 CFU/g。微囊化益生菌在乳制品中应用6周内下降值为$10^{1.16}$ CFU/g	de Castro-Cislaghi et al., 2012
热变性分离乳清蛋白质溶液10%（w/w）	短双歧杆菌R070和长双歧杆菌R023	喷雾干燥技术，进风和出风温度分别为160℃和80℃	短双歧杆菌R070包埋率：25.67%±0.12%。长双歧杆菌R023包埋率：1.44%±0.16%。	Picot and Lacroix, 2004
乳清蛋白质和抗性淀粉	鼠李糖乳杆菌GG	喷雾干燥技术，进风和出风温度分别为160℃和65℃	圆形小的微粒	Ying et al., 2010
浓缩乳清蛋白质：12%（w/w）	长双歧杆菌1941	冻干	含有不同糖的浓缩乳清蛋白质微胶囊中益生菌存活率：80%~95%	Dianawati et al., 2013
变性分离乳清蛋白质溶液	鼠李糖乳杆菌	挤压技术	微粒直径：约3 mm，包埋效率：96%，存活率：23%	Ainsley et al. 2005

1）喷雾干燥法

喷雾干燥法是食品工业中最实用的方法。在喷雾干燥过程中，一些疏水性益生菌会与乳清蛋白质暴露的疏水性区域产生疏水相互作用，从而将益生菌包埋在胶囊壁内（Khem et al., 2016）。喷雾干燥过程中低存活率及后期储藏过程低稳定性是喷雾干燥工艺的一个弊端。喷雾干燥工艺条件的优化要同时考虑喷雾效率和益生菌存活率。通过优化喷雾干燥条件（如入口和出口温度）并在干燥前添加保护剂可获得较高存活率（Ananta et al., 2005）。鼠李糖乳杆菌GG的喷雾干燥微囊化过程中，使用乳清蛋白质和抗性淀粉作为壁材可提高益生菌存活率。与冷冻干燥微囊化技术相比，保护剂的使用将导致更高的生存效率和更好的储存稳定性（Ying et al., 2010）。

2）冷冻干燥法

冷冻干燥更适合益生菌微囊化。Dianawati等（2013）比较了五种蛋白质和三种糖类化合物在冷冻干燥微囊化过程中对长双歧杆菌的保护作用,结果表明浓缩乳清蛋白质/酪蛋白酸钠与甘油在冷冻(99.2%)和冷冻干燥(97.2%)过程中对益生菌的稳定性保护作用最强。

3)挤出技术

多糖是益生菌微囊化常用壁材,然而,乳清蛋白质/多糖复合物通常被用作复合壁材,以提高益生菌存活率,因为β-LG 能够抵抗胃蛋白酶水解作用(Gbassi et al., 2009)。在离子诱导条件下,乳清蛋白质可形成冷凝胶。这些冷凝胶微粒是耐胃酸分解的。因此,乳清蛋白质的冷凝胶化是一种包埋热敏性活性物质的有效方法。

凝胶微粒的形成包括两个过程:①预变性(加热)步骤,即通过加热的方法使蛋白质暴露疏水位点进而形成聚集体。②通过注射器针头或微胶囊机将含有益生菌的变性蛋白溶液挤出到氯化钙溶液中,固化形成微胶囊颗粒。Hébrard 等用乳清蛋白质溶液(10%)包埋酿酒酵母,形成了直径约 1 mm 的凝胶珠。在胃液条件下,微囊化酿酒酵母的活性明显高于游离酵母(Hébrard et al., 2006)。乳清蛋白质基鼠李糖乳杆菌凝胶微胶囊直径约为 3 mm,在模拟胃液条件下,凝胶微粒可为鼠李糖乳杆菌提供长达 90 min 的保护,并在十二指肠具有长达 90 min 的抗胆汁能力(Ainsley et al., 2005)。凝胶化乳清蛋白质微粒包埋的鼠李糖乳杆菌 GG 在 pH 分别为 3.4、2.4 和 2.0 孵育 180 min 后,存活率分别提高了 $10^{5.7}$ CFU/mL、$10^{5.1}$ CFU/mL 和 $10^{2.2}$ CFU/mL,而游离细胞在模拟胃液(pH 3.4)条件下反应 30 min 后,存活率为零(Doherty et al., 2012)。

多糖和蛋白质可通过静电或表面疏水性相互作用作为复合壁材。在以果胶为壁材的微胶囊中,果胶颗粒表面的负电荷允许蛋白质通过静电相互作用吸附。当蛋白质添加量为 4%时,天然乳清蛋白质(干基 49.2%)比热处理乳清蛋白质(干基为 27.6%)显示出更高的吸附果胶性能(Souza et al., 2012)。吸附蛋白层的果胶微胶囊可作为益生菌载体并应用于低 pH 的功能性食品(如苹果汁),从而保护鼠李糖乳杆菌 CRL 1505 免受胃部酸性环境影响(Gerez et al., 2012)。海藻酸钙微粒吸附乳清蛋白质包埋益生菌也显示出了更高的益生菌存活能力。尽管海藻酸盐和乳清蛋白质在中性 pH 上都带有负电荷,但由于乳清蛋白质具有较高表面疏水性,因此其可形成一层壁材。干酪乳杆菌在溶液中带负电荷,因此能够与海藻酸盐相互作用(Smilkov et al., 2014)。植物乳杆菌通过静电相互作用被 WPI 和κ-卡拉胶复合物包埋,并且在低 pH 和胆汁盐条件下表现出更好的存活率(Hernandez-Rodriguez et al., 2014)。

7.5.3 微囊化益生菌在食品中的应用

微囊化益生菌被广泛应用于食品中。微胶囊添加对食物的感官特性以及在食品中益生菌稳定性的影响被广泛研究。Picot 和 Lacroix 以 10%热变性 WPI 溶液为壁材,通过喷雾干燥微囊化短双歧杆菌 R070,并将其应用在酸奶中。在 4℃下放置 4 周后,封装后短双歧杆菌 R070 数量仅下降 $10^{2.5}$ CFU/g,而未封装的下降了 $10^{5.1}$ CFU/g(Picot and Lacroix,2004)。以果胶/乳清蛋白质复合物为壁材,采用离子凝胶法封装嗜酸乳杆菌 LA-5,并将其应用在酸奶中。与游离的益生菌相比,添加微囊化益生菌的酸奶在冷藏 35 天后,有较低的酸化程度和较高的益生菌存活率。在储存结束时,添加游离益生菌和微囊化益生菌的酸奶样品在外观、香气、风味和整体接受度方面无显著性差异(Ribeiro et al.,2014)。

7.5.4 生物活性成分的微囊化

近年来,随着人们对健康的关注,生物活性成分受到了越来越多的关注。但是,某些特性限制了一些生物活性成分的广泛应用,如人参皂苷的苦味和颜色(Wang et al.,2017),花青素的降解敏感性(Betz and Kulozik,2011),白藜芦醇的低溶解性、光和热敏感性(Lee et al.,2013)。微囊化技术能够克服很多活性物质弊端,从而将其更广泛应用于食品。

Wang 等利用乳清蛋白质的冷凝胶特性制备微囊化人参皂苷,随后应用于发酵乳中,结果表明含有微囊化人参皂苷的发酵乳制品比含有人参皂苷提取物的样品具有更高的可接受性。此外,微囊化人参皂苷显著降低了发酵乳样品的脱水收缩率(Wang et al.,2017)。来源于越橘、蓝莓和石榴的花青素已被乳清蛋白质微囊化。利用乳液法制备的乳清蛋白质基花青素微胶囊(Betz and Kulozik,2011)与未封装的越橘花青素相比,在 pH 6.8 时花青素的降解受到抑制(Betz et al.,2012)。

以 WPI 为壁材,采用喷雾干燥法微囊化蓝莓果渣提取物,微胶囊在整个消化过程中释放率低,但抗氧化活性高(Flores et al.,2014b)。以戊二醛饱和的甲苯(GAST)交联的乳清蛋白质为壁材包埋茶碱,微胶囊芯材含量和芯材保留率分别是 6.7%~65.7%和 16.8%~85.4%,具体取决于芯材和壁材的比例范围(1:1.5~5:1.5)(Lee and Rosenberg,2000)。采用乳清蛋白质的复合壁材包埋茶碱,壁材含有不同比例分散的非极性填充剂,无水乳脂通过戊二醛饱和的甲苯有机相交联。芯材含量为 46.9%~56.6%,微囊化过程中保留率为 84.9%~96.9%(Lee and Rosenberg,2001)。以 WPI 和可溶性玉米纤维作为复合壁材,采用乳化法和喷雾干燥法包埋虾青素,微囊化效率高达 95%。微囊化粉末具有相当好的性能,包括水分活度、表面形态、微囊化效率和氧化稳定性(Shen and Quek,2014)。

维生素 A(又称视黄醇)是一种疏水性物质,对胃部低 pH 环境敏感。视黄醇乳液是通过混合变性乳清蛋白质和含有视黄醇的大豆油,然后高压均质化得到。再将乳液滴加到 $CaCl_2$ 溶液中形成微胶囊。微胶囊具有抵抗胃肠降解,并在肠道靶向吸收的特点（Beaulieu et al., 2002）。维生素 D_3 被包埋在通过不同钙浓度制备的分离乳清蛋白质纳米颗粒中，颗粒成分中存在钙会导致形成致密结构并抑制颗粒中的氧扩散（Abbasi et al., 2014）。

除了视黄醇之外，胆盐水解酶等其他敏感性物质的递送也值得关注，这些物质在通过较低 pH 的胃部环境的递送过程中可能会对蛋白酶的活性造成伤害，以乳清蛋白质/阿拉伯胶为壁材制备的胆盐水解酶微胶囊已证实能在胃部环境下有效保护其活性并能实现肠部定点释放（Lambert et al., 2008）。

7.6 食用膜和涂层

食用膜和涂层能够提供一种对气体和水蒸气的半渗透性屏障，从而减少食物呼吸和水分流失，降低氧化反应速率。乳清蛋白质基食用膜和涂层的机械性能及阻隔性能已得到广泛研究。

7.6.1 食用膜和涂层参数

1. 膜的机械性能

根据标准方法 ASTMD882-91 测定膜的拉伸强度（TS）、弹性模量（EM）和伸长率（E, %）。

2. 膜/涂层厚度

膜的厚度可以通过游标卡尺测量。涂层可从食物表面上剥离，然后确定。

3. 膜溶解度

膜在水中的溶解度是指浸入蒸馏水中 24 h 后溶解的薄膜干物质百分比。将一部分膜置于蒸馏水中并搅拌约 1 天，然后通过真空过滤并干燥。浸渍过程中干物质的损失就是溶解在水中的薄膜。溶解度通常表示为干物质损失相对于初始干物质的百分比。

4. 渗透性

包装材料的渗透性对于维持包装产品的高质量至关重要。食品成分的氧化会导致其产生异味、变色和营养损失，因此避免食品氧化是评价包装材料好坏的重要指标之一（Ramos et al., 2012a）。氧气渗透率（OP）可根据美国材料与试验协会（D3985/D618）标准方法测量。透水性（WP）是评估乳清蛋白质基薄膜的重要指标。但乳清蛋白质是亲水性蛋白，具有适度的水分屏障作用。透水性通常通过杯法确定，将乳清蛋白质基薄膜安装在塑料杯上并密封，在一定的湿度和温度下放入干燥器中后于一定时间间隔内称量杯子，对质量和时间进行线性回归分析，线的斜率即为水蒸气透过率（WVTR）。水蒸气渗透率（WVP）根据式（7.6）计算。

$$WVP = \frac{WVTR \times 厚度}{WVPP} \quad (7.6)$$

其中，WVTR 以 g/(h·m²) 为单位；厚度以 mm 为单位；WVPP 是水蒸气分压，分压以 kPa 为单位。

5. 氧化和变色

乳清蛋白质涂层通常用于保护高油高脂食物（主要由不饱和脂肪酸组成）以及水果。通常情况下，涂层有效性可直接通过感官评价食品在存储过程中的氧化和颜色变化来进行评估。氧化还可以通过测量过氧化物值（PV）或利用顶空气相色谱法测定己醛含量来评估，颜色的变化可通过色度计测量。

7.6.2 乳清蛋白质薄膜/涂层

图 7.4 为制备乳清蛋白质薄膜/涂层的一般流程图。首先，将 WPC 或 WPI（5%～12%）溶解在去离子水中，将乳清蛋白质溶液调节至 pH 7 或 8 或不调节 pH，然后在 80～90℃下加热 10～30 min，形成变性乳清蛋白质。紫外处理（Diaz et al., 2017）以及转谷氨酰胺酶交联乳清蛋白质（Marquez et al., 2017）制备变性蛋白或聚集乳清蛋白质进而制备膜溶液，可以在加热之前或之后添加增塑剂。通常，在该步骤中添加抗氧化剂、抗菌剂或益生菌等。乳清蛋白质食用膜是通过将乳清蛋白质膜溶液导入平板中，并在一定条件下干燥制备而得。对于涂层，通常将食品（包括水果、海鲜和肉类）浸入膜溶液中 30 s 到 1 min，以确保涂层能很好地铺满食品整个表面。大多数乳清蛋白质基可食用薄膜是将薄膜溶液铺在板上后通过空气干燥制备而得。温度、湿度和干燥速率影响干燥过程以及膜的特性。将含有蜂蜡的乳清蛋白质乳液薄膜在 5℃和 25℃两个温度条件下干燥。将干燥温度从

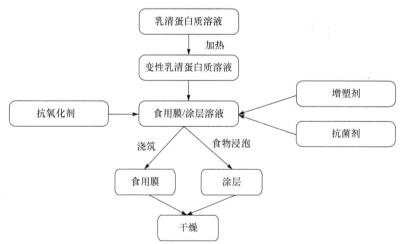

图 7.4 乳清蛋白质基食用膜/涂层制备流程图

25℃降低至5℃可降低水蒸气渗透性并增加薄膜溶解度（Soazo et al.，2011）。使用微波干燥乳清蛋白质可食用薄膜仅需 5 min，而在室温条件下空气干燥需要 18 h。微波干燥薄膜的水蒸气渗透率与室温干燥方法相似，但微波干燥增加了食用膜的伸长率和拉伸强度值（Kaya and Kaya，2000）。不同干燥温度和湿度的组合会显著影响薄膜性能。在 95℃和 30%RH 下干燥的薄膜比在 21℃和 50%RH 下干燥的薄膜具有更低的水蒸气渗透性。与在 21℃和 50%RH 下干燥的薄膜相比，在 95℃和 30%RH 下干燥的薄膜通常更薄、更硬、更坚固且延展性更差（Alcantara et al.，2010）。

7.6.3 乳清蛋白质可食用膜/涂层的组成

1. 蛋白质类型和浓度

变性乳清蛋白质已被广泛用作食用膜基质。对于天然乳清蛋白质基食用膜和涂层，氢键起主要作用，而对于变性乳清蛋白质食用膜，二硫键和氢键均有助于形成食用膜/涂层。天然乳清蛋白质膜完全溶于水，而热变性乳清蛋白质膜则不溶于水。变性乳清蛋白质膜的溶解度随着加热时间和温度的升高而降低（Perez-Gago and Krochta，2001）。热变性乳清蛋白质薄膜比天然乳清蛋白质薄膜具有更高的拉伸性能。随着成膜时间和温度的增加，薄膜变得更硬、更坚固且更具可拉伸性（Perez-Gago and Krochta，2001）。天然乳清蛋白质膜和热变性乳清蛋白质膜表现出相似的水蒸气渗透性（Perez-Gago etal.，1999）。热变性乳清蛋白质制成的薄膜氧渗透性（OP）低于天然乳清蛋白质制成的薄膜（Perez-Gago and Krochta，2001）。

WPI 和 WPC 涂层在阻氧性能方面未观察到明显差异（Hong and Krochta，2006）。加热（90℃、30 min）乳清蛋白质膜和β-LG 膜，在水蒸气渗透性和氧渗透性方面无显著性差异。甘油作为增塑剂可能掩盖了两种材料之间的某些差异。同时，β-LG 与其他乳清蛋白质组分以相似的方式促进了 WPI 基质的屏障性能（Maté and Krochta，1996）。

乳清蛋白质和增塑剂的含量对所形成的膜的渗透性有影响。将 5%、7%和 9%的 WPI 与甘油（Gly）混合，比例分别为 3.6∶1、3∶1 和 2∶1，WPI∶Gly 为 3.6∶1 的 WPI（5%）表现出最佳的水蒸气透过率，而 WPI∶Gly 为 3.6∶1 的 9% WPI 显示出最佳的氧气透过率（Gounga et al.，2007）。在由蛋白质、山梨糖醇、蜂蜡和山梨酸钾复合而成的薄膜中，蛋白质浓度影响薄膜的水蒸气渗透性和水溶性，但是对黏性和外观没有影响（Ozdemir and Floros，2008b）。蛋白质浓度会影响拉伸强度、杨氏模量和伸长率（Ozdemir and Floros，2008a）。

2. 增塑剂类型和浓度

由于乳清蛋白质基膜成膜中的主要作用力因素，乳清蛋白质膜比较脆弱（Schmid et al.，2013）。通常情况下，可通过添加增塑剂提高弹性并避免乳清蛋白质膜破裂（Sothornvit and Krochta，2000）。增塑剂为一种不易挥发、高沸点、不易分离的物质，当添加到另一种材料中时会改变该材料的物理和/或机械性能（Banker，1966）。增塑剂的添加减少了蛋白质与蛋白质的相互作用并增加了聚合物链的迁移率。然而，纯乳清蛋白质膜中的强交联也是其具有优异阻隔性能的原因。增塑剂的添加还导致蛋白质网络中更高的自由体积，从而增加了膜的渗透性，这在食品应用中是不希望的（Schmid et al.，2013）。增塑剂对乳清蛋白质基食用膜综合物理性能的影响取决于增塑剂的类型和浓度。山梨糖醇和甘油是乳清蛋白质基薄膜中常用的增塑剂，同时添加两种增塑剂会显著影响薄膜性能。将10%的乳清蛋白质溶液在90℃加热30 min，在冷却的溶液中加入相同质量的山梨糖醇和甘油，由甘油浓度增加引起的透氧性增加大于山梨糖醇。对于具有相同拉伸强度、伸长率和弹性模量的食用膜来说，使用山梨糖醇作为增塑剂的食用膜比使用甘油作为增塑剂的食用膜氧气渗透性更低（Mchugh and Krochta，1994）。

食用膜中增塑剂的迁移会导致乳清蛋白质基薄膜出现一定程度的不稳定性。在食用膜储藏期间，与山梨醇相比，甘油是一种更有效的增塑剂，用甘油增塑的膜不影响膜的外观或机械性能。此外，与使用相同量的山梨糖醇相比甘油增塑的膜更柔软，更有韧性并且含水量更高。这可能是由于山梨糖醇结晶，山梨糖醇增塑膜在储藏期间随着时间的变化逐渐变硬，柔韧性变低（Javier et al.，2009）。

山梨糖醇（俗称葡萄糖醇）是一种糖醇，具有甜味，可被人体缓慢代谢。它可以通过还原葡萄糖醛基成羟基而获得。山梨糖醇对透湿性及薄膜的水溶性具有一定影响。山梨糖醇的含量（35%~50%）对食用膜外观没有影响（Ozdemir and Floros，2008b）。除了对食用膜溶解性和含水量有影响外（Kim and Ustunol，2001），山梨糖醇也影响食用膜拉伸强度、杨氏模量和伸长率（Ozdemir and Floros，2008a）。

甘油是一种简单的多元醇化合物。它是一种无色、无气味、黏稠的液体，味甜，无毒。食品工业中，甘油被广泛用作甜味剂和保湿剂，以及应用于药物制剂配方中。甘油溶于水，富有吸湿性主要是因为甘油分子中含有三个羟基。甘油添加量影响乳清蛋白质基食用膜的溶解度和平衡水分含量（Kim and Ustunol，2001）。增加甘油含量能够提高乳清蛋白质膜的溶解度，降低表观杨氏模量和玻璃化转变温度（Galietta et al.，1998）。乳清蛋白质基食用膜的平衡水分含量随甘油浓度线性增加（Coupland et al.，2000）。

3. 脂质

乳清蛋白质膜的亲水性使其在水分屏障方面效果较差。在食用膜中掺入蜂蜡等有效屏障水分的脂质物质，可大大改善这些薄膜的水分屏障能力（Soazo et al.，2011）。

4. 防腐剂

为延长食用膜保质期并改善食用膜和涂层固有的保护性能，在食用膜和涂层制备过程中经常添加一些防腐剂，如山梨糖醇和山梨酸钾。食用膜中山梨糖醇含量（35%~50%）对薄膜外观无影响（Ozdemir and Floros，2008b），但山梨糖醇和山梨酸钾是影响极限拉伸强度、杨氏模量和薄膜伸长率的重要因素（Ozdemir and Floros，2008a）。

5. 抑菌剂

抑菌包装是一种活性包装方法，可以延长产品的保质期并为消费者提供微生物安全性。抑菌包装是目前食品安全领域最受欢迎的话题之一。植物基化合物如牛至精油、迷迭香、大蒜精油、肉桂精油、对氨基苯甲酸等经常被用作活性食用膜制备中的抑菌剂。

牛至精油因其较好的抗氧化和抑菌活性，已成为食品中应用最广泛的精油（Oliveira et al.，2017）。与含大蒜提取物和迷迭香提取物的食用膜相比，研究人员发现掺入牛至精油（含量为2%）的乳清蛋白质膜对大肠杆菌O157:H7（ATCC 35218）、金黄色葡萄球菌（ATCC 43300）、肠炎沙门氏菌（ATCC 13076）、单核细胞增生李斯特菌（NCTC 2167）和植物乳杆菌（DSM 20174）有更好的抑菌性能（Seydim and Sarikus，2006）。研究人员还制备了含有对氨基苯甲酸（PABA）或山梨酸（SA）的pH 5.2的WPI可食用膜。添加1.5% PABA和1.5% SA的食用膜对单核细胞增生李斯特菌、大肠杆菌O157:H7和鼠伤寒沙门氏菌DT104的平均抑制圈直径分别为21.8 mm、14.6 mm、13.9 mm和26.7 mm、10.5 mm、9.7 mm（Cagri et al.，2001）。含肉桂精油乳清蛋白质食用膜对革兰氏阳性和革兰氏阴性菌株均显示出明显的抑菌活性（Bahram et al.，2014）。土耳其法兰克福香肠表面涂有乳链菌肽、苹果酸、葡萄籽提取物和乙二胺四乙酸（EDTA），在4℃储存28天后能有效抑制单核细胞增生李斯特菌、鼠伤寒沙门氏菌以及大肠杆菌O157:H7（Gadang et al.，2008）。含有牛至精油的WPC食用膜抑制了paínhos和alheiras的微生物含量，并分别延长了两种产品的保存期限约20天和15天（Catarino et al.，2017）。含有3%和4%的大蒜精油的乳清蛋白质基可食膜对大肠杆菌O157:H7（ATCC 35218）、金黄色葡萄球菌（ATCC 43300）、肠炎沙门氏菌（ATCC 13076）、单核细胞增生李斯特菌（NCTC 2167）和植物乳杆菌（DSM 20174）均有抑制活性（Seydim and Sarikus，2006）。抑菌剂的添加不仅能够抑制微生物，还能够改变食用膜的物理性质。

牛至精油过能影响乳清蛋白质膜的物理性能（制备条件为蛋白浓度7%，90℃加热30 min）。随着牛至精油浓度（0.5%~1.5%）的增加，食用膜弹性增大，透湿性增高，水溶性下降。添加1%牛至精油提高了乳清蛋白质膜的耐水性。含有1%牛至精油的食用膜的拉伸强度为108.7 MPa，远远高于对照组食用膜的强度（66.0 MPa）（Oliveira et al.，2017）。对氨基苯甲酸和山梨酸的添加提高了食用膜的伸长率，但降低了拉伸强度。0.5%和0.75%的山梨酸添加量对食用膜透湿性无影响,但添加对氨基苯甲酸提高了食用膜的透湿性(Cagri et al., 2001)。

6. 益生菌

在食品中添加益生菌能够使食品功能化,并改善食品的健康功效(Soukoulis et al., 2017)。添加益生菌的乳清蛋白质基薄膜也因其抑菌活性而被应用。含有清酒乳杆菌 NRRL B-1917 的乳清蛋白质食用膜冷藏 120 h 后,单核细胞增生李斯特菌降低了 $10^{1.4}$ CFU/g,储藏 36 h 后大肠杆菌降低了 $10^{2.3}$ CFU/g(Beristainbauza et al., 2017)。

7.6.4 乳清蛋白质/多糖复合膜的物理性质

不同的蛋白质和多糖制备而成的食用膜具有不同的理化性质。乳清蛋白质具有良好的氧、气味和油阻隔性能,然而,由于其氨基酸基团的亲水性,它对水蒸气的渗透性较高。蛋白质和多糖(如藻酸盐、果胶和羧甲基纤维素)之间的静电复合物能够结合蛋白质和多糖的性能,从而改善蛋白质基食用膜的性能(Sabato et al., 2001)。羟丙基甲基纤维素(HPMC)/乳清蛋白质膜比单独乳清蛋白质膜更坚固。

在 WPI 薄膜中添加 HPMC 能够影响食用膜的柔韧性、强度、拉伸能力和水溶性(Brindle and Krochta, 2008)。HPMC/乳清蛋白质复合物食用膜比单独乳清蛋白质膜具有较低的透氧性(Yoo and Krochta, 2011)。淀粉和分离乳清蛋白质复合物可用于制备可食用膜。可食用膜表面性质取决于淀粉/乳清蛋白质比例。由 80%淀粉和 20%乳清蛋白质组成的食用膜比其他食用膜有更多的疏水表面(Basiak et al., 2017)。

7.6.5 乳清蛋白质膜在食品工业中的应用

与非食用膜相比,由食品级材料制成的薄膜可以食用,具有一定的竞争优势。

1. 干酪

干酪是一种即食食品,表面很容易被微生物污染。含有抑菌剂的可食用乳清蛋白质基涂层的应用可抑制不良微生物生长,从而延长保质期。添加纳他霉素、乳酸和壳寡糖不同组合的抑菌剂的乳清蛋白质食用膜的干酪在储藏过程中乳酸菌能够正常生长,但抑制致病菌或污染微生物生长(Ramos et al., 2012a)。干酪涂有壳聚糖/乳清蛋白质可食膜后,与未包装产品相比,在储藏过程中,嗜温和嗜冷微生物含量明显降低(Pierro et al., 2011)。添加乳链菌肽、苹果酸和那他霉素浸渍的乳清蛋白质膜在干酪表面上能有效抑制铜绿假单胞菌、解脂耶氏酵母、单核细胞增生李斯特菌、青霉菌和产黄青霉活性(Pintado et al., 2010)。在阻止水分渗透和减少质量方面,干酪上的可食用抗菌涂层可与商业化涂层相媲美(Ramos et al., 2012b)。在干酪上涂上含有纳他霉素、乳酸、壳寡糖等几种抑菌化合物组合的食用膜，可减少水分损失（约 10%），减少硬度和颜色变化（Ramos et al., 2012a）。

2. 鸡蛋和肉

鸡蛋在储藏过程中会因蛋白、蛋黄、质量和 pH 变化而迅速丧失品质。这些变化主要是鸡蛋中二氧化碳和水分通过蛋壳上的细孔发生质量转移而引起的。乳清蛋白质食用膜可在室温下有效延长新鲜鸡蛋保质期。乳清蛋白质膜包被的鸡蛋在室温下存储约 4 周时的质量损失为 2.38%~2.46%，而未涂膜的鸡蛋为 5.66%。涂膜鸡蛋比未涂膜鸡蛋表现出更好的蛋白品质和更低的 pH。涂膜鸡蛋卵黄指数为 0.26~0.9，比未涂膜鸡蛋更接近优质鸡蛋卵黄指数（Caner，2010）。

加有天然抗氧化剂的乳清蛋白质食用膜已成功应用于香肠和熟肉丸中。与对照相比，涂有乳清蛋白质膜的低脂香肠在有氧包装下于 4℃储存 8 周后显示出较低的硫代巴比妥酸活性产物（TBARS）和过氧化值（PV）以及水分流失（Shon and Chin，2008）。含有月桂或鼠尾草的天然抗氧化剂提取物的分离乳清蛋白质食用膜可在-18℃冷冻储存 60 天期间抑制煮熟肉丸的氧化，降低对氨基苯甲酸值和硫代巴比妥酸值（Akcan et al.，2017）。含有乳链菌肽、葡萄籽提取物、苹果酸的乳清蛋白质膜可在 4℃下储藏 28 天时，将土耳其法兰克福香肠单核细胞增生李斯特菌种群由 $10^{5.5}$ CFU/g 减少至 $10^{2.3}$ CFU/g，并对鼠伤寒沙门氏菌和大肠杆菌 O157:H7 有良好抑菌活性（Gadang et al.，2008）。

3. 海鲜

由于高持水性、中性 pH、含组织酶、结缔组织含量低和天然微生物污染等，鱼是一种易腐食品（Rodriguez-Turienzo et al.，2011）。冷冻是保存鱼类的重要方法。然而，冷冻过程中经常会发生一些不良变化，如蛋白质变性、质量减轻和脂质氧化。在海鲜保藏中可使用磷酸盐增强保水能力，使用丁基羟基茴香醚（BHA）和丁基羟基甲苯（BHT）抑制脂质氧化，从而延长海鲜的货架期（Sathivel，2005）。随着消费者食品安全意识的增强，可生物降解的蛋白质（如乳清蛋白质涂层）可用于涂层鱼片，以抑制冷冻储藏过程中可能存在的质量变化。三文鱼脂肪含量高，并且主要是不饱和脂肪酸，极易氧化。在 3 周储藏期内，乳清蛋白质膜涂层三文鱼可使水分流失率降低 42%~65%，并延迟脂质氧化，降低过氧化值（Stuchell and Krochta，2010）。涂层时间对冷冻效果具有一定的影响，与冷冻前涂层相比，冷冻后涂层增加了三文鱼融化率，减少了汁液损失，改变了冷冻和解冻鱼片的颜色（Rodriguez-Turienzo et al.，2011）。随后，同一研究团队利用 8% WPC 或 WPI 溶液制备了涂层溶液，并添加了甘油作为增塑剂，在 35 kHz 的频率下进行了超声波处理（1 min、15 min 或 60 min）。所有乳清蛋白质涂层均能延迟三文鱼鱼块的脂质氧化，而不会影响样品感官特性（Rodriguez-Turienzo et al.，2012）。

4. 水果

由于内源多酚氧化酶与酚类化合物的酶促反应以及空气中氧气向组织中扩散作用，鲜切水果经常发生酶促褐变。乳清蛋白质涂层可通过屏障氧气来延迟褐变。乳清蛋白质比酪蛋白酸钙具有更好的抗氧化能力（Le Tien et al., 2001）。涂有 WPI 或 WPC 和蜂蜡或巴西棕榈蜡食用膜的苹果片具有较高的 L^* 和较低的 b^*、a^* 值，表明乳清蛋白质涂层具有抗褐变作用（Perez-Gago et al., 2005）。WPC 和蜂蜡基涂层（含 1%抗坏血酸或 0.5%半胱氨酸）实际上是防止苹果褐变的最有效方法（Perez-Gago et al., 2006）。WPI 涂层的性能取决于环境相对湿度。随着相对湿度的降低，内部氧气减少，二氧化碳增加，在低相对湿度（70%~80%相对湿度）下，由于氧气含量低（大约 0.025 atm①），带涂层水果被诱导产生无氧呼吸（Cisneros-Zevallos and Krochta, 2003）。添加羧甲基纤维素可显著提高乳清蛋白质抗氧化能力，可抑制氧化性物质与 N,N-二乙基对苯二甲胺反应生成有色化合物，抑制率达 75%（Le Tien et al., 2001）。乳清蛋白质食用膜已广泛应用于苹果上，还应用于李子、冷冻干燥的板栗和草莓片上。与未涂膜李子比较，乳清蛋白质包膜的李子接受度更高，水分损失更少（Reinoso et al., 2008）。涂有乳清蛋白质/普鲁兰多糖膜的新鲜烤制和冷冻干燥的板栗在储藏过程中均延迟外部颜色变化（Gounga et al., 2008）。WPI 涂层显著降低了冻干草莓片在牛奶中的复水率，解决了补水速度过快和冻干草莓质地松弛的问题（Huang et al., 2009）。

5. 坚果

花生油中不饱和脂肪酸含量高，在焙烤高温条件下容易发生氧化酸败。脂质氧化是花生变质的主要原因之一。包装或可食用涂层可降低脂质氧化速率。乳清蛋白质基涂层可通过屏障氧气而提供保护机制（Maté et al., 1996）。研究表明，天然或热变性的分离乳清蛋白质（添加或不添加维生素）涂层均能延迟花生氧化，己醛分析结果表明乳清蛋白质膜在 40℃、50℃和 60℃条件下将花生的保质期延长至 31 周（Lee and Krochta, 2002）。通过固相微萃取和气相色谱-质谱仪（GC-MS）测定的无盐烤花生储藏过程中己醛含量，结果表明 WPI 食用膜可延迟干烤花生氧化降解。值得注意的是，较大的厚度和较低的相对湿度会使涂膜效果更好（Han et al., 2008）。含抗坏血酸的 WPI 涂层在 23℃、35℃和 50℃时显著抑制了花生脂质氧化，并且涂层花生比未涂层花生更红（Min and Krochta, 2007）。在干烤花生上应用乳清蛋白质涂层时，无论是否加入抗氧化剂，都会延迟氧化性酸败。然而，研究证明尽管 WPI 涂层的氧气渗透率低且连续性好，但并不能明显延迟核桃的氧化酸败（Maté and Krochta, 1996），可能的原因是乳清蛋白质涂层在干燥过程中收缩导致一些油从核桃中被挤压出来。

---

① 1 atm = 1.01325×10^5 Pa。

7.7 总　　结

乳清蛋白质或聚合乳清蛋白质具有良好的功能特性，已广泛应用于食品工业中，用来改善食品的质地和质量。本章基于乳清蛋白质的凝胶特性、乳化特性、结合疏水性物质特性、成膜特性，综述了其作为增稠剂/胶凝剂、脂肪替代品、乳化剂/稳定剂、疏水性营养食品载体、微胶囊壁材、食用膜和涂层的应用。尽管已有大量的相关研究，但仍需开展优化乳清蛋白质相关食品配方和工艺相关研究。同时，需要进一步阐明乳清蛋白质在食品中发挥作用的相关机制。

参 考 文 献

Abbasi, A., Emam-Djomeh, Z., Mousavi, M.A.E. et al. (2014). Stability of vitamin D-3 encapsulated in nanoparticles of whey protein isolate. *Food Chemistry* **143**: 379-383.

Ainsley, R.A., Vuillemard, J.C., Britten, M. et al. (2005). Microentrapment of probiotic bacteria in a Ca^{2+}-induced whey protein gel and effects on their viability in a dynamic gastro-intestinal model. *Journal of Microencapsulation* **22** (6): 603-619.

Akcan, T., Estevez, M., and Serdaroglu, M. (2017). Antioxidant protection of cooked meatballs during frozen storage by whey protein edible films with phytochemicals from Laurus nobilis L. and Salvia officinalis. *Lwt-Food Science and Technology* **77**: 323-331.

Alcantara, C.R., Rumsey, T.R., and Krochta, J.M. (2010). Drying rate effect on the properties of whey protein films. *Journal of Food Process Engineering* **21** (5): 387-405.

Alting, A.C., van der Meulena, E.T., Hugenholtz, J. et al. (2004). Control of texture of cold-set gels through programmed bacterial acidification. *International Dairy Journal* **14** (4): 323-329.

Anal, A.K. and Singh, H. (2007). Recent advances in microencapsulation of probiotics for industrial applications and targeted delivery. *Trends in Food Science and Technology* **18** (5): 240-251.

Ananta, E., Volkert, M., and Knorr, D. (2005). Cellular injuries and storage stability of spray-dried *Lactobacillus rhamnosus* GG. *International Dairy Journal* **15** (4): 399-409.

Arslan, S., Erbas, M., Tontul, I. et al. (2015). Microencapsulation of probiotic *Saccharomyces cerevisiae* var. boulardii with different wall materials by spray drying. *Lwt-Food Science and Technology* **63** (1): 685-690.

Bahram, S., Rezaei, M., Soltani, M. et al. (2014). Whey protein concentrate edible film activated with cinnamon essential oil. *Journal of Food Processing and Preservation* **38** (3): 1251-1258.

Banker, G.S. (1966). Film coating theory and practice. *Journal of Pharmaceutical Sciences* **55** (1): 81-89.

Basiak, E., Lenart, A., and Debeaufort, F. (2017). Effects of carbohydrate/protein ratio on the microstructure and the barrier and sorption properties of wheat starch-whey protein blend edible films. *Journal of the Science of Food and Agriculture* **97** (3): 858-867.

Beaulieu, L., Savoie, L., Paquin, P. et al. (2002). Elaboration and characterization of whey protein beads by an emulsification/cold gelation process: application for the protection of retinol. *Biomacromolecules* **3** (2): 239-248.

Beristainbauza, S.D., Manilópez, E., Palou, E. et al. (2017). Antimicrobial activity of whey protein films supplemented with *Lactobacillus sakei* cell-free supernatant on fresh beef. *Food Microbiology* **62**: 207-211.

Betz, M. and Kulozik, U. (2011). Microencapsulation of bioactive bilberry anthocyanins by means of whey protein gels. *Procedia Food Science* **1**: 2047-2056.

Betz, M., Steiner, B., Schantz, M. et al. (2012). Antioxidant capacity of bilberry extract microencapsulated in whey protein hydrogels. *Food Research International* **47** (1): 51-57.

Borgogna, M., Bellich, B., Zorzin, L. et al. (2010). Food microencapsulation of bioactive compounds: rheological and thermal characterisation of non-conventional gelling system. *Food Chemistry* **122** (2): 416-423.

Brindle, L.P. and Krochta, J.M. (2008). Physical properties of whey protein-hydroxypropyl methylcellulose blend edible films. *Journal of Food Science* **73** (9): 446-454.

Bryant, C.M. and Mcclements, D.J. (1998). Molecular basis of protein functionality with special consideration of cold-set gels derived from heat-denatured whey. *Trends in Food Science and Technology* **9** (4): 143-151.

Burgar, M.I., Hoobin, P., Weerakkody, R. et al. (2009). NMR of microencapsulated fish oil samples during in vitro digestion. *Food Biophysics* **4** (1): 32-41.

Bylaite, E., Venskutonis, P.R., and Mapdpieriene, R. (2001). Properties of caraway (Carum carvi L.) essential oil encapsulated into milk protein-based matrices. *European Food Research and Technology* **212** (6): 661-670.

Cagri, A., Ustunol, Z., and Ryser, E.T. (2001). Antimicrobial, mechanical, and moisture barrier properties of low pH whey protein-based edible films containing p-aminobenzoic or sorbic acids. *Journal of Food Science* **66** (6): 865-870.

Caner, C. (2010). Whey protein isolate coating and concentration effects on egg shelf life. *Journal of the Science of Food and Agriculture* **85** (13): 2143-2148.

Carneiro, H.C.F., Tonon, R.V., Grosso, C.R.F. et al. (2013). Encapsulation efficiency and oxidative stability of flaxseed oil microencapsulated by spray drying using different combinations of wall materials. *Journal of Food Engineering* **115** (4): 443-451.

Catarino, M.D., Alves-Silva, J.M., Fernandes, R.P. et al. (2017). Development and performance of whey protein active coatings with origanum virens essential oils in the quality and shelf life improvement of processed meat products. *Food Control* **80**: 273-280.

Charoen, R., Jangchud, A., Jangchud, K. et al. (2011). Influence of biopolymer emulsifier type on formation and stability of rice bran oil-in-water emulsions: whey protein, gum arabic, and modified starch. *Journal of Food Science* **76** (1): E165-E172.

Cheng, J.J., Xie, S.Y., Yin, Y. et al. (2017). Physiochemical, texture properties, and the microstructure of set yogurt using whey protein-sodium tripolyphosphate aggregates as thickening agents. *Journal of the Science of Food and Agriculture* **97** (9): 2819-2825.

Cho, Y.H., Shim, H.K., and Park, J. (2003). Encapsulation of fish oil by an enzymatic gelation process using transglutaminase cross-linked proteins. *Journal of Food Science* **68** (9): 2717-2723.

Chung, C., Degner, B., and McClements, D.J. (2014). Development of reduced-calorie foods: microparticulated whey proteins as fat mimetics in semi-solid food emulsions. *Food Research International* **56**: 136-145.

Chung, C., Rojanasasithara, T., Mutilangi, W. et al. (2015). Enhanced stability of anthocyanin-based color in model beverage systems through whey protein isolate complexation. *Food Research International* **76**: 761-768.

Cisneros-Zevallos, L. and Krochta, J.M. (2003). Whey protein coatings for fresh fruits and relative humidity effects. *Journal of Food Science* **68** (1): 176-181.

Coupland, J.N., Shaw, N.B., Monahan, F.J. et al. (2000). Modeling the effect of glycerol on the moisture sorption behavior of whey protein edible films. *Journal of Food Engineering* **43** (1): 25-30.

Damodaran, S. (2005). Protein stabilization of emulsions and foams. *Journal of Food Science* **70** (3): R54-R66.

Damodaran, S. and Paraf, A. (1997). Protein-stablized foams and emulsions. In: *Food Proteins and Their Applications* (ed. S. Damodaran and A. Paraf), 57-111. New York: Marcel Dekker.

De Castro-Cislaghi, F.P., Silva, C.D.E., Fritzen-Freire, C.B. et al. (2012). *Bifidobacterium* Bb-12 microencapsulated by spray drying with whey: survival under simulated gastrointestinal conditions, tolerance to NaCl, and viability during storage. *Journal of Food Engineering* **113** (2): 186-193.

Demetriades, K., Coupland, J.N., and Mcclements, D.J. (2010). Physicochemical properties of whey protein-stabilized emulsions as affected by heating and ionic strength. *Journal of Food Science* **62** (3): 462-467.

Desai, K.G.H. and Park, H.J. (2005). Recent developments in microencapsulation of food ingredients. *Drying Technology* **23** (7): 1361-1394.

Desrumaux, A. and Marcand, J. (2002). Formation of sunflower oil emulsions stabilized by whey proteins with high-pressure homogenization (up to 350 MPa): effect of pressure on emulsion characteristics. *International Journal of Food Science and Technology* **37** (3): 263-269.

Dufour, E. and Haertlé, T. (1991). Binding of retinoids and β-carotene to β-lactoglobulin. Influence of protein modifications. *Biochimica et Biophysica Acta (BBA)-Protein Structure and Molecular Enzymology* **1079** (3): 316-320.

Dianawati, D., Mishra, V., and Shah, N.P. (2013). Survival of *Bifidobacterium longum* 1941 microencapsulated with proteins and sugars after freezing and freeze drying. *Food Research International* **51** (2): 503-509.

Diaz, O., Candia, D., and Cobos, A. (2017). Whey protein film properties as affected by ultraviolet treatment under alkaline conditions. *International Dairy Journal* **73**: 84-91.

Dickinson, E. (1994). Protein-stabilized emulsions. In: *Water in Foods: Fundamental Aspects and Their Significance in Relation to Processing of Foods* (ed. P. Fito, A. Mulet and B. McKenna), 57-94. Oxford: Pergamon Press.

Dickinson, E. and Euston, S.R. (1991). Stability of food emulsions containing both protein and polysaccharide. In: *Food Polymers, Gels and Colloids* (ed. E. Dickinson), 132-146. Cambridge: The Royal Society of Chemistry.

Doherty, S.B., Auty, M.A., Stanton, C. et al. (2012). Survival of entrapped *Lactobacillus rhamnosus* GG in whey protein micro-beads during simulated ex vivo gastro-intestinal transit. *International Dairy Journal* **22** (1): 31-43.

Elofsson, C., Dejmek, P., Paulsson, M. et al. (1997). Characterization of a cold-gelling whey protein concentrate. *International Dairy Journal* **7** (8-9): 601-608.

Fioramonti, S.A., Martinez, M.J., Pilosof, A.M.R. et al. (2015). Multilayer emulsions as a strategy for linseed oil microencapsulation: effect of pH and alginate concentration. *Food Hydrocolloids* **43**: 8-17.

Fitzsimons, S.M., Mulvihill, D.M., and Morris, E.R. (2008). Large enhancements in thermogelation of whey protein isolate by incorporation of very low concentrations of guar gum. *Food Hydrocolloids* **22** (4): 576-586.

Flores, F.P., Singh, R.K., Kerr, W.L. et al. (2014a). Total phenolics content and antioxidant capacities of microencapsulated blueberry anthocyanins during in vitro digestion. *Food Chemistry* **153**: 272-278.

Flores, F.P., Singh, R.K., and Kong, F.B. (2014b). Physical and storage properties of spray-dried blueberry pomace extract with whey protein isolate as wall material. *Journal of Food Engineering* **137**: 1-6.

Gadang, V.P., Hettiarachchy, N.S., Johnson, M.G. et al. (2008). Evaluation of antibacterial activity of whey protein isolate coating incorporated with nisin, grape seed extract, malic acid, and EDTA on a Turkey frankfurter system. *Journal of Food Science* **73** (8): M389-M394.

Galietta, G., Gioia, L.D., Guilbert, S. et al. (1998). Mechanical and thermomechanical properties of films based on whey proteins as affected by plasticizer and crosslinking agents. *Journal of Dairy Science* **81** (12): 3123-3130.

Gallardo, G., Guida, L., Martinez, V. et al. (2013). Microencapsulation of linseed oil by spray drying for functional food application. *Food Research International* **52** (2): 473-482.

Gbassi, G.K., Vandamme, T., Ennahar, S. et al. (2009). Microencapsulation of *Lactobacillus plantarum* spp in an alginate matrix coated with whey proteins. *International Journal of Food Microbiology* **129** (1): 103-105.

Ge, J., Yue, P.X., Chi, J.P. et al. (2018). Formation and stability of anthocyanins-loaded nanocomplexes prepared with chitosan hydrochloride and carboxymethyl chitosan. *Food Hydrocolloids* **74**: 23-31.

Gerez, C.L., de Valdez, G.F., Gigante, M.L. et al. (2012). Whey protein coating bead improves the survival of the probiotic *Lactobacillus rhamnosus* CRL 1505 to low pH. *Letters in Applied Microbiology* **54** (6): 552-556.

Gounga, M.E., Xu, S.Y., and Wang, Z. (2007). Whey protein isolate-based edible films as affected by protein concentration, glycerol ratio and pullulan addition in film formation. *Journal of Food Engineering* **83** (4): 521-530.

Gounga, M.E., Xu, S.Y., Wang, Z. et al. (2008). Effect of whey protein isolate-pullulan edible coatings on the quality and shelf life of freshly roasted and freeze-dried Chinese chestnut. *Journal of Food Science* **73** (4): E155-E161.

Hébrard, G., Blanquet, S., Beyssac, E. et al. (2006). Use of whey protein beads as a new carrier system for recombinant yeasts in human digestive tract. *Journal of Biotechnology* **127** (1): 151-160.

Han, J.H., Hwang, H.M., Min, S. et al. (2008). Coating of peanuts with edible whey protein film containing alpha-tocopherol and ascorbyl palmitate. *Journal of Food Science* **73** (8): E349-E355.

Hebishy, E., Zamora, A., Buffa, M. et al. (2017). Characterization of whey protein oil-in-water emulsions with different oil concentrations stabilized by ultra-high pressure homogenization. *Processes* **5** (1): 6-24.

Henriques, M., Gomes, D., Rodrigues, D. et al. (2011). Performance of bovine and ovine liquid whey protein concentrate on functional properties of set yoghurts. *Procedia Food Science* **1**: 2007-2014.

Hernandez-Rodriguez, L., Lobato-Calleros, C., Pimentel-Gonzalez, D.J. et al. (2014). *Lactobacillus plantarum* protection by entrapment in whey protein isolate: kappa-carrageenan complex coacervates. *Food Hydrocolloids* **36**: 181-188.

Hogan, S.A., McNamee, B.F., O'Riordan, E.D. et al. (2001). Microencapsulating properties of whey protein concentrate 75. *Journal of Food Science* **66** (5): 675-680.

Hong, S.I. and Krochta, J.M. (2006). Oxygen barrier performance of whey-protein-coated plastic films as affected by temperature, relative humidity, base film and protein type. *Journal of Food Engineering* **77** (3): 739-745.

Huang, L.L., Zhang, M., Yan, W.Q. et al. (2009). Effect of coating on post-drying of freeze-dried strawberry pieces. *Journal of Food Engineering* **92** (1): 107-111.

Hundre, S.Y., Karthik, P., and Anandharamakrishnan, C. (2015). Effect of whey protein isolate and beta-cyclodextrin wall systems on stability of microencapsulated vanillin by spray-freeze drying method. *Food Chemistry* **174**: 16-24.

Hunt, J.A. and Dalgleish, D.G. (1994). Adsorption behaviour of whey protein isolate and caseinate in soya oil-in-water emulsions. *Food Hydrocolloids* **8** (2): 175-187.

Ibanoglu, E. (2002). Rheological behaviour of whey protein stabilized emulsions in the presence of gum arabic. *Journal of Food Engineering* **52** (3): 273-277.

Implvo, F., Pinho, O., Mota, M.V. et al. (2007). Preparation of ingredients containing an ACEinhibitory peptide by tryptic hydrolysis of whey protein concentrates. *International Dairy Journal* **17** (5): 481-487.

Jafari, S.M., Assadpoor, E., He, Y.H. et al. (2008). Encapsulation efficiency of food flavours and oils during spray drying. *Drying Technology* **26** (7): 816-835.

Javier, O.S., Idoya, F.N., Mauricio, M. et al. (2009). Stability of the mechanical properties of edible films based on whey protein isolate during storage at different relative humidity. *Food Hydrocolloids* **23** (1): 125-131.

Jimenez, M., Garcia, H.S., and Beristain, C.I. (2004). Spray-drying microencapsulation and oxidative stability of conjugated linoleic acid. *European Food Research and Technology* **219** (6): 588-592.

Maté, J.I. and Krochta, J.M. (1997). Whey protein and acetylated monoglyceride edible coatings:

effect on the rancidity process of walnuts. *Journal of Agricultural and Food Chemistry* **45** (7): 2509-2513.

Kavas, G., Oysun, G., Kinik, O. et al. (2004). Effect of some fat replacers on chemical, physical and sensory attributes of low-fat white pickled cheese. *Food Chemistry* **88** (3): 381-388.

Kaya, S. and Kaya, A. (2000). Microwave drying effects on properties of whey protein isolate edible films. *Journal of Food Engineering* **43** (2): 91-96.

Khem, S., Small, D.M., and May, B.K. (2016). The behaviour of whey protein isolate in protecting *Lactobacillus plantarum*. *Food Chemistry* **190**: 717-723.

Kim, S.J., Cho, S.Y., Kim, S.H. et al. (2008). Effect of microencapsulation on viability and other characteristics in *Lactobacillus acidophilus* ATCC 43121. *Lwt-Food Science and Technology* **41** (3): 493-500.

Kim, S.J. and Ustunol, Z. (2001). Solubility and moisture sorption isotherms of whey-protein-based edible films as influenced by lipid and plasticizer incorporation. *Journal of Agricultural and Food Chemistry* **49** (9): 4388-4391.

Kiokias, S., Dimakou, C., and Oreopoulou, V. (2007). Effect of heat treatment and droplet size on the oxidative stability of whey protein emulsions. *Food Chemistry* **105** (1): 94-100.

Klein, M., Aserin, A., Svitov, I. et al. (2010). Enhanced stabilization of cloudy emulsions with gum Arabic and whey protein isolate. *Colloids and Surfaces B-Biointerfaces* **77** (1): 75-81.

Kraan, S. and Dominguez, H. (2013). Pigments and minor compounds in algae. In: *Functional Ingredients from Algae for Foods and Nutraceuticals* (ed. H. Dominguez), 205-251. UK: Wood head.

Kuhn, K.R. and Cunha, R.L. (2012). Flaxseed oil-whey protein isolate emulsions: effect of high pressure homogenization. *Journal of Food Engineering* **111** (2): 449-457.

Kulmyrzaev, A., Chanamai, R., and McClements, D.J. (2000). Influence of pH and CaCl2 on the stability of dilute whey protein stabilized emulsions. *Food Research International* **33** (1): 15-20.

Kulmyrzaev, A.A. and Schubert, H. (2004). Influence of KCl on the physicochemical properties of whey protein stabilized emulsions. *Food Hydrocolloids* **18** (1): 13-19.

Lambert, J.M., Weinbreck, F., and Kleerebezem, M. (2008). In vitro analysis of protection of the enzyme bile salt hydrolase against enteric conditions by whey protein-gum arabic microencapsulation. *Journal of Agricultural and Food Chemistry* **56** (18): 8360-8364.

Le Tien, C., Vachon, C., Mateescu, M.A. et al. (2001). Milk protein coatings prevent oxidative browning of apples and potatoes. *Journal of Food Science* **66** (4): 512-516.

Lee, S.J. and Rosenberg, M. (2000). Microencapsulation of theophylline in whey proteins: effects of core-to-wall ratio. *International Journal of Pharmaceutics* **205** (1-2): 147-158.

Lee, S.J. and Rosenberg, M. (2001). Microencapsulation of theophylline in composite wall system consisting of whey proteins and lipids. *Journal of Microencapsulation* **18** (3): 309-321.

Lee, S.Y. and Krochta, J.M. (2002). Accelerated shelf life testing of whey-protein-coated peanuts analyzed by static headspace gas chromatography. *Journal of Agricultural and Food Chemistry* **50** (7): 2022-2028.

Lee, Y.K., Ahn, S.I., and Kwak, H.S. (2013). Optimizing microencapsulation of peanut sprout extract by response surface methodology. *Food Hydrocolloids* **30** (1): 307-314.

Liu, D., Zhang, T., Jiang, N. et al. (2016). Effects of encapsulated fish oil by polymerized whey protein on the textural and sensory characteristics of low-fat yogurt. *Polish Journal of Food and Nutrition Sciences* **66** (3): 189-198.

Lyons, P.H., Kerry, J.F., Morrissey, P.A. et al. (1999). The influence of added whey protein/carrageenan gels and tapioca starch on the textural properties of low fat pork sausages. *Meat Science* **51** (1): 43-52.

Mandal, S., Puniya, A.K., and Singh, K. (2006). Effect of alginate concentrations on survival of microencapsulated *Lactobacillus casei* NCDC-298. *International Dairy Journal* **16** (10): 1190-1195.

Mantovani, R.A., Pinheiro, A.C., Vicente, A.A. et al. (2017). In vitro digestion of oil-in-water emulsions stabilized by whey protein nanofibrils. *Food Research International* **99**: 790-798.

Marquez, G.R., Di Pierro, P., Mariniello, L. et al. (2017). Fresh-cut fruit and vegetable coatings by transglutaminase-crosslinked whey protein/pectin edible films. *Lwt-Food Science and Technology* **75**: 124-130.

Maté, J.I. and Krochta, J.M. (1996). Comparison of oxygen and water vapor permeabilities of whey protein isolate and beta-lactoglobulin edible films. *Journal of Agriculture and Food Chemistry* **44** (10): 3001-3004.

Maté, J.I., Frankel, E.N., and Krochta, J.M. (1996). Whey protein isolate edible coatings: effect on the rancidity process of dry roasted peanuts. *Journal of Agriculture and Food Chemistry* **44** (7): 1736-1740.

Mchugh, T.H. and Krochta, J.M. (1994). Sorbitol-vs glycerol-plasticized whey protein edible films: integrated oxygen permeability and tensile property evaluation. *Journal of Agricultural and Food Chemistry* **42** (4): 841-845.

Miguel, M.G. (2011). Anthocyanins: antioxidant and/or anti-inflammatory activities. *Journal of Applied Pharmaceutical Science* **1** (6): 7-15.

Min, S. and Krochta, J.M. (2007). Ascorbic acid-containing whey protein film coatings for control of oxidation. *Journal of Agricultural and Food Chemistry* **55** (8): 2964-2969.

Mirpoor, S.F., Hosseini, S.M.H., and Yousefi, G.H. (2017). Mixed biopolymer nanocomplexes conferred physicochemical stability and sustained release behavior to introduced curcumin. *Food Hydrocolloids* **71**: 216-224.

Moschakis, T., Murray, B.S., and Biliaderis, C.G. (2010). Modifications in stability and structure of whey protein-coated o/w emulsions by interacting chitosan and gum arabic mixed dispersions. *Food Hydrocolloids* **24** (1): 8-17.

Na, H.S., Kim, J.N., Kim, J.M. et al. (2011). Encapsulation of fish oil using cyclodextrin and whey protein concentrate. *Biotechnology and Bioprocess Engineering* **16** (6): 1077-1082.

Neirynck, N., Meeren, P.V.D., Lukaszewicz-Lausecker, M. et al. (2007). Influence of pH and biopolymer ratio on whey protein-pectin interactions in aqueous solutions and in O/W emulsions. *Colloids & Surfaces A Physicochemical & Engineering Aspects* **298** (1): 99-107.

Neirynck, N., Van der Meeren, P., Gorbe, S.B. et al. (2004). Improved emulsion stabilizing properties of whey protein isolate by conjugation with pectins. *Food Hydrocolloids* **18** (6): 949-957.

Ng, S.K., Nyam, K.L., Lai, O.M. et al. (2017). Development of a palm olein oil-in-water (o/w) emulsion stabilized by a whey protein isolate nanofibrils-alginate complex. *Lwt-Food Science and Technology* **82**: 311-317.

O'Brien, C.M., Mueller, A., Scannell, A.G.M. et al. (2003). Evaluation of the effects of fat replacers on the quality of wheat bread. *Journal of Food Engineering* **56** (2-3): 265-267.

Oliveira, S.P.L.F., Bertan, L.C., Bilck, A.P. et al. (2017). Whey protein-based films incorporated with oregano essential oil. *Polímeros* **27** (2): 158-164.

Ozdemir, M. and Floros, J.D. (2008a). Optimization of edible whey protein films containing preservatives for mechanical and optical properties. *Journal of Food Engineering* **84** (1): 116-123.

Ozdemir, M. and Floros, J.D. (2008b). Optimization of edible whey protein films containing preservatives for water vapor permeability, water solubility and sensory characteristics. *Journal of Food Engineering* **86** (2): 215-224.

Partanen, R., Raula, J., Seppanen, R. et al. (2008). Effect of relative humidity on oxidation of flaxseed oil in spray dried whey protein emulsions. *Journal of Agricultural and Food Chemistry* **56** (14): 5717-5722.

Perez-Gago, M.B. and Krochta, J.M. (2001). Denaturation time and temperature effects on solubility, tensile properties, and oxygen permeability of whey protein edible films. *Journal of Food Science* **66** (5): 705-710.

Perez-Gago, M.B., Nadaud, P., and Krochta, J.M. (1999). Water vapor permeability, solubility, and tensile properties of heat-denatured versus native whey protein films. *Journal of Food Science* **64** (6): 1034-1037.

Perez-Gago, M.B., Serra, M., Alonso, M. et al. (2005). Effect of whey protein- and hydroxypropyl methylcellulose-based edible composite coatings on color change of fresh-cut apples. *Postharvest Biology and Technology* **36** (1): 77-85.

Perez-Gago, M.B., Serra, M., and del Rio, M.A. (2006). Color change of fresh-cut apples coated with whey protein concentrate-based edible coatings. *Postharvest Biology and Technology* **39** (1): 84-92.

Picot, A. and Lacroix, C. (2004). Encapsulation of bifidobacteria in whey protein-based microcapsules and survival in simulated gastrointestinal conditions and in yoghurt. *International Dairy Journal* **14** (6): 505-515.

Pierro, P.D., Sorrentino, A., Mariniello, L. et al. (2011). Chitosan/whey protein film as active coating to extend ricotta cheese shelf-life. *LWT-Food Science and Technology* **44** (10): 2324-2327.

Pintado, C.M.B.S., Ferreira, M.A.S.S., and Sousa, I. (2010). Control of pathogenic and spoilage microorganisms from cheese surface by whey protein films containing malic acid, nisin and natamycin. *Food Control* **21** (3): 240-246.

Ramos, Ó.L., Pereira, J.O., Silva, S.I. et al. (2012a). Evaluation of antimicrobial edible coatings from a whey protein isolate base to improve the shelf life of cheese. *Journal of Dairy Science* **95** (11): 6282-6292.

Ramos, Ó.L., Fernandes, J.C., Silva, S.I. et al. (2012b). Edible films and coatings from whey proteins:

a review on formulation, and on mechanical and bioactive properties. *Critical Reviews in Food Science and Nutrition* **52** (6): 533-552.

Ramos, Ó.L., Santos, A.C., Leão, M.V. et al. (2012c). Antimicrobial activity of edible coatings prepared from whey protein isolate and formulated with various antimicrobial agents. *International Dairy Journal* **25** (2): 132-141.

Reinoso, E., Mittal, G.S., and Lim, L.T. (2008). Influence of whey protein composite coatings on plum (Prunus Domestica L.) fruit quality. *Food and Bioprocess Technology* **1** (4): 314-325.

Ribeiro, H.S., Guerrero, J.M., Briviba, K. et al. (2006). Cellular uptake of carotenoid-loaded oil-inwater emulsions in colon carcinoma cells in vitro. *Journal of Agricuture and Food Chemistry* **54** (25): 9366-9369.

Ribeiro, H.S., Rico, L.G., Badolato, G.G. et al. (2005). Production of O/W emulsions containing astaxanthin by repeated premix membrane emulsification. *Journal of Food Science* **70** (2): E117-E123.

Ribeiro, M.C.E., Chaves, K.S., Gebara, C. et al. (2014). Effect of microencapsulation of *Lactobacillus acidophilus* LA-5 on physicochemical, sensory and microbiological characteristics of stirred probiotic yoghurt. *Food Research International* **66**: 424-431.

Rodriguez-Turienzo, L., Cobos, A., and Diaz, O. (2012). Effects of edible coatings based on ultrasoundtreated whey proteins in quality attributes of frozen Atlantic salmon (Salmo salar). *Innovative Food Science & Emerging Technologies* **14**: 92-98.

Rodriguez-Turienzo, L., Cobos, A., Moreno, V. et al. (2011). Whey protein-based coatings on frozen Atlantic salmon (Salmo solar): influence of the plasticiser and the moment of coating on quality preservation. *Food Chemistry* **128** (1): 187-194.

Rosenberg, M. and Sheu, T.Y. (1996). Microencapsulation of volatiles by spray-drying in whey proteinbased wall systems. *International Dairy Journal* **6** (3): 273-284.

Sabato, S.F., Ouattara, B., Yu, H. et al. (2001). Mechanical and barrier properties of cross-linked soy and whey protein based films. *Journal of Agricultural & Food Chemistry* **49** (3): 1397-1403.

Sakandar, H.A., Imran, M., Huma, N. et al. (2014). Effects of polymerized whey proteins isolates on the quality of stirred yoghurt made from camel milk. *Journal of Food Processing & Technology* **5** (7): https://doi.org/10.4172/2157-7110.1000350.

Sathivel, S. (2005). Chitosan and protein coatings affect yield, moisture loss, and lipid oxidation of pink salmon (Oncorhynchus gorbuscha) fillets during frozen storage. *Journal of Food Science* **70** (8): E455-E459.

Schmid, M., Hinz, L.V., Wild, F. et al. (2013). Effects of hydrolysed whey proteins on the technofunctional characteristics of whey protein-based films. *Materials* **6** (3): 927-940.

Setiowati, A.D., Saeedi, S., Wijaya, W. et al. (2017). Improved heat stability of whey protein isolate stabilized emulsions via dry heat treatment of WPI and low methoxyl pectin: effect of pectin concentration, pH, and ionic strength. *Food Hydrocolloids* **63**: 716-726.

Seydim, A.C. and Sarikus, G. (2006). Antimicrobial activity of whey protein based edible films incorporated with oregano, rosemary and garlic essential oils. *Food Research International* **39** (5): 639-644.

Shen, Q. and Quek, S.Y. (2014). Microencapsulation of astaxanthin with blends of milk protein and fiber by spray drying. *Journal of Food Engineering* **123**: 165-171.

Shen, X., Zhao, C., Lu, J. et al. (2018). Physicochemical properties of whey protein-stabilized astaxanthin nanodispersion and its transport via Caco-2 monolayer. *Journal of Agricultural and Food Chemistry* **66** (6): 1472-1478.

Shon, J. and Chin, K.B. (2008). Effect of whey protein coating on quality attributes of low-fat, aerobically packaged sausage during refrigerated storage. *Journal of Food Science* **73** (6): C469-C475.

Stuchell, Y.M. and Krochta, J.M. (2010). Edible coatings on frozen king salmon: effect of whey protein isolate and acetylated monoglycerides on moisture loss and lipid oxidation. *Journal of Food Science* **60** (1): 28-31.

Sliwinski, E.L., Roubos, P.J., Zoet, F.D. et al. (2003). Effects of heat on physicochemical properties of whey protein-stabilised emulsions. *Colloids and Surfaces B-Biointerfaces* **31** (1-4): 231-242.

Smilkov, K., Ivanovska, T.P., Tozi, L.P. et al. (2014). Optimization of the formulation for preparing *Lactobacillus casei* loaded whey protein-Ca-alginate microparticles using full-factorial design. *Journal of Microencapsulation* **31** (2): 166-175.

Soazo, M., Rubiolo, A.C., and Verdini, R.A. (2011). Effect of drying temperature and beeswax content on physical properties of whey protein emulsion films. *Food Hydrocolloids* **25** (5): 1251-1255.

Sothornvit, R. and Krochta, J.M. (2000). Oxygen permeability and mechanical properties of films from hydrolyzed whey protein. *Journal of Agricultural and Food Chemistry* **48** (9): 3913-3916.

Soukoulis, C., Behboudijobbehdar, S., Macnaughtan, W. et al. (2017). Stability of *Lactobacillus rhamnosus* GG incorporated in edible films: impact of anionic biopolymers and whey protein concentrate. *Food Hydrocolloids* **70**: 345-355.

Souza, F.N., Gebara, C., Ribeiro, M.C.E. et al. (2012). Production and characterization of microparticles containing pectin and whey proteins. *Food Research International* **49** (1): 560-566.

Sun, C.H. and Gunasekaran, S. (2009). Effects of protein concentration and oil-phase volume fraction on the stability and rheology of menhaden oil-in-water emulsions stabilized by whey protein isolate with xanthan gum. *Food Hydrocolloids* **23** (1): 165-174.

Sun, C.H., Gunasekaran, S., and Richards, M.P. (2007). Effect of xanthan gum on physicochemical properties of whey protein isolate stabilized oil-in-water emulsions. *Food Hydrocolloids* **21** (4): 555-564.

Tonon, R.V., Pedro, R.B., Grosso, C.R.F. et al. (2012). Microencapsulation of flaxseed oil by spray drying: effect of oil load and type of wall material. *Drying Technology* **30** (13): 1491-1501.

Vardhanabhuti, B., Foegeding, E.A., McGuffey, M.K. et al. (2001). Gelation properties of dispersions containing polymerized and native whey protein isolate. *Food Hydrocolloids* **15** (2): 165-175.

Visentini, F.F., Sponton, O.E., and Perez, A.A. (2016). Formation and colloidal stability of ovalbuminretinol nanocomplexes. *Food Hydrocolloids* **67**: 130-138.

Walsh, H., Ross, J., Hendricks, G. et al. (2010). Physico-chemical properties, probiotic survivability, microstructure, and acceptability of a yogurt-like symbiotic oats-based product using pre-polymerized wheyprotein as a gelation agent. *Journal of Food Science* **75** (5): M327-M337.

Wang, C.N., Gao, F., Zhang, T.H. et al. (2015). Physiochemical, textural, sensory properties and probiotic survivability of Chinese Laosuan Nai (protein-fortified set yoghurt) using polymerised whey protein as a co-thickening agent. *International Journal of Dairy Technology* **68** (2): 261-269.

Wang, C.N., Zheng, H.J., Liu, T.T. et al. (2017a). Physiochemical properties and probiotic survivability of symbiotic corn-based yogurt-like product. *Journal of Food Science* **82** (9): 2142-2150.

Wang, M., Gao, F., Zheng, H.J. et al. (2017b). Microencapsulation of ginsenosides using polymerized whey protein (PWP) as wall material and its application in probiotic fermented milk. *International Journal of Food Science and Technology* **52** (4): 1009-1017.

Wang, S.J., Shi, Y., Tu, Z.C. et al. (2017c). Influence of soy lecithin concentration on the physical properties of whey protein isolate-stabilized emulsion and microcapsule formation. *Journal of Food Engineering* **207**: 73-80.

Wang, W.B., Bao, Y.H., Hendricks, G.M. et al. (2012). Consistency, microstructure and probiotic survivability of goats' milk yoghurt using polymerized whey protein as a co-thickening agent. *International Dairy Journal* **24** (2): 113-119.

Xu, D.X., Wang, X.Y., Jiang, J.P. et al. (2012). Impact of whey protein-beet pectin conjugation on the physicochemical stability of beta-carotene emulsions. *Food Hydrocolloid* **28** (2): 258-266.

Yan, B., Park, S.H., and Balasubramaniam, V.M. (2017). Influence of high pressure homogenization with and without lecithin on particle size and physicochemical properties of whey protein-based emulsions. *Journal of Food Process Engineering* **40** (6): https://doi.org/10.1111/jfpe.12578.

Ye, A. and Singh, H. (2000). Influence of calcium chloride addition on the properties of emulsions stabilized by whey protein concentrate. *Food Hydrocolloids* **14** (4): 337-346.

Ye, A.Q. (2008). Interfacial composition and stability of emulsions made with mixtures of commercial sodium caseinate and whey protein concentrate. *Food Chemistry* **110** (4): 946-952.

Ying, D.Y., Phoon, M.C., Sanguansri, L. et al. (2010). Microencapsulated *Lactobacillus rhamnosus* GG powders: relationship of powder physical properties to probiotic survival during storage. *Journal of Food Science* **75** (9): E588-E595.

Yoo, S. and Krochta, J.M. (2011). Whey protein-polysaccharide blended edible film formation and barrier, tensile, thermal and transparency properties. *Journal of the Science of Food and Agriculture* **91** (14): 2628-2636.

Zhang, S. and Vardhanabhuti, B. (2014). Effect of initial protein concentration and pH on in vitro gastric digestion of heated whey proteins. *Food Chemistry* **145**: 473-480.

Zhang, T.H., McCarthy, J., Wang, G.R. et al. (2015). Physiochemical properties, microstructure, and probiotic survivability of nonfat goats' milk yogurt using heat-treated whey protein concentrate as fat replacer. *Journal of Food Science* **80** (4): M788-M794.

Zhu, J., Sun, X., Wang, S. et al. (2017). Formation of nanocomplexes comprising whey proteins and fucoxanthin: characterization, spectroscopic analysis, and molecular docking. *Food Hydrocolloids* **63**: 391-403.

第 8 章 乳清蛋白质的改性

Mingruo Guo[1,2] and Xue Shen[3]

1. Department of Nutrition and Food Science, University of Vermont, Burlington, USA
2. College of Food Science, Northeast Agriculture University, Harbin, People's Republic of China
3. Department of Food Science, College of Food Science and Engineering, Jilin University, ChangChun, People's Republic of China

乳清蛋白质具有较高的营养价值和一定的功能特性，如凝胶性、乳化性、起泡性、风味结合特性等，已被广泛应用于许多传统和新型食品中（Smithers，2015）。这些功能特性有助于形成各种食品的结构、感官和营养特性。乳清蛋白质的改性是改变或增强其物理性和功能性的必要条件，这可能会产生新的蛋白质并扩大其应用范围。乳清蛋白质的理化和功能性质可以通过热诱导聚合（热处理）(Ryan and Foegeding，2015）、酶处理（Wang et al.，2013b）、化学改性（Leila et al.，2019）以及通过高压、极化电场（PEF）、γ射线和超声波等其他新技术得以改变。这些技术为扩大乳清蛋白质在食品和非食品领域中的应用提供了新的机会。

8.1 热 处 理

热处理是食品加工中最常用的方法之一，对乳清蛋白质的理化和功能特性有很大的影响。虽然严重的热处理会降低乳清蛋白质的营养价值，但适度加热可以改善乳清蛋白质的乳化性、起泡能力和热性能（Kim et al.，2005；Nicolai et al.，2011；Schmitt et al.，2007）。乳清蛋白质经热改性后表面疏水性增加，这是因为在加热和暴露初始嵌入的疏水性氨基酸残基时，分子发生了皱缩（Hussain et al.，2012）。乳清蛋白质经热处理后黏度增加。

热处理乳清蛋白质的功能特性直接依赖于乳清蛋白质聚合物的结构。乳清蛋白质可溶性聚合物主要是指加热过程中形成的蛋白质单体与不溶性凝胶网络或沉淀物之间的中间产物（Ryan et al.，2013）。它们也被称为聚合乳清蛋白质、乳清蛋白质纳米颗粒或预热乳清蛋白质。当乳清蛋白质受热时，通过二硫键交换和键合，乳清蛋白质发生解离和去折叠，形成中等大小的聚合物，然后通过非共价键相互作用形成可溶性聚合物。当 pH>5.7 时，乳清蛋白质热诱导聚合过程的机理如图 8.1 所示。

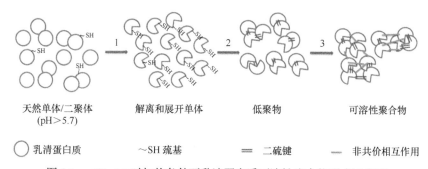

图 8.1　pH>5.7 时加热条件下乳清蛋白质可溶性聚合物形成示意图

天然乳清蛋白质（单体/二聚体）在加热过程中展开并分解成单体。暴露活性巯基基团并形成反应单体（步骤 1）。这些反应单体通过二硫键相互作用形成低聚物，主要是二聚体、三聚体和其他较小的聚合物（步骤 2）。这些低聚物通过共价和非共价相互作用形成更大的可溶性聚合物（步骤 3）。在适当的条件下（pH、离子类型和浓度、温度及蛋白质浓度）进一步加热，聚合物可能会沉淀或形成凝胶。

热诱导的乳清蛋白质聚合物的流体力学直径可达 300 nm（Guo and Wang, 2016）。浓度为 5%～15% 的乳清蛋白质（pH 6～8）在 55～120℃加热 1～120 min，可获得乳清蛋白质可溶性聚合物（Wijayanti et al., 2014）。可溶性聚合物的形成与处理条件有关，包括温度、时间、剪切速率、pH、离子类型和浓度以及蛋白质浓度（Nicolai et al., 2011）。通过对热处理条件的控制，可以获得具有预期理化性质(如表面电荷、表面疏水性)和具有不同大小的球形颗粒以及分形簇（fractal clusters）、柔性链（flexible strands）和长半柔性纤维（long semi-flexible）等广泛形态的乳清蛋白质聚合物（Nicolai and Durand, 2013）。这些特性反过来又可用于获得目标的宏观特性，如黏度和最终聚合物的起泡性。

在较高的蛋白质浓度下，乳清蛋白质凝胶是通过初级聚合物进一步聚集形成的。乳清蛋白质凝胶基于其流变学和微观结构可划分为细链凝胶、混合凝胶和颗粒凝胶。乳清蛋白质凝胶形成示意图如图 8.2 所示。乳清蛋白质可以形成不同大小和形状的初级聚合物，如柔性链或球状物。这些初级聚合物进一步聚集（也称为次级聚合）形成凝胶网络。静电条件（pH 和离子强度）和蛋白质浓度对乳清蛋白质的凝胶特性有很大的影响。在天然乳清蛋白质溶液中添加乳清蛋白质可溶性聚合物，可以提高凝胶的断裂应力、储能模量、保水能力和热致凝胶的透明度（Vardhanabhuti et al., 2001）。

可以优化乳清蛋白质的热改性以控制聚合物的特性，这为扩大乳清蛋白质的应用提供了新的机会。热致乳清蛋白质聚合物可作为一种新型的蛋白质基增稠剂，用于酸奶等产品的生产。同时，乳清蛋白质聚合物也可以作为低脂酸奶或低脂干酪的脂肪替代品。热致乳清蛋白质聚合物具有良好的黏结性能，可广泛应用于环

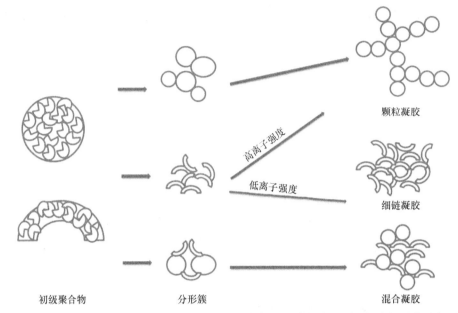

图 8.2 加热过程中乳清蛋白质凝胶（包括细链凝胶、混合凝胶和颗粒凝胶）形成示意图
当分子间有静电斥力时（pH 远离 pI，离子强度低），通过柔性链的进一步集聚，可以形成细链凝胶。颗粒凝胶可以在等电点附近形成，也可以通过进一步聚集球形颗粒形成二价离子。颗粒凝胶也可以在高离子强度下通过柔性链的进一步集聚而形成

境安全型胶黏剂中。此外，乳清蛋白质聚合物还可用于生产可食性薄膜或包衣，以延长货架期和提高食品品质。另外，乳清蛋白质可溶性聚合物可以通过加盐或降低 pH 来制备冷凝凝胶。乳清蛋白质聚合物形成的网络也被用作保护和运输敏感食品组分和生物活性化合物的一种包埋体系。乳清蛋白质聚合物作为一种良好的乳化剂和稳定剂，可以用来稳定乳浊液，有助于泡沫的稳定。

8.2 酶 处 理

酶常用于处理各种原料，乳清蛋白质的酶改性可以通过共价交联或水解来实现，不仅可以提高乳清蛋白质的功能性，而且可以提高乳清蛋白质的生物活性（Eallen and Jackp，2011）。因此，酶改性在食品工业中得到了广泛的应用。

8.2.1 转谷氨酰胺酶交联

转谷氨酰胺酶（TGase）广泛存在于自然界所有的动植物和微生物细胞中。TGase 能在较宽的 pH 范围内（pH 5~8）保持较高的活性。TGase 在 50~55℃表现出最佳活性。TGase（酶代码 EC 2.3.2.13，蛋白质-谷氨酰胺-γ-谷氨酰转移酶）通过在蛋白质分子内或分子间ε-(γ-谷氨酰基)-赖氨酸之间形成共价异肽键来催化

酰基转移反应（Kuraishi et al.，2001）。TGase 可通过胺融合、交联和脱氨基作用对蛋白质进行修饰（图 8.3）。蛋白质经酶处理后，会产生高分子量聚合物，导致蛋白质功能性质的改变。

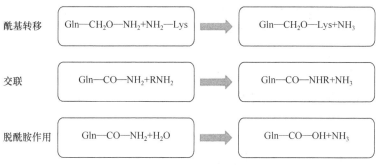

图 8.3　谷氨酰胺转胺酶催化反应

改编自 Jaros et al.（2006）

β-乳球蛋白（β-LG）和α-乳清蛋白质（α-LA）是乳清蛋白质中的主要成分，它们是致密的球状蛋白。由于其具有球状结构，乳清蛋白质不易与 TGase 发生交联反应。因此，在暴露酶靶位点（即反应性赖氨酸和谷氨酰胺分子）的情况下展开蛋白质分子可以促进酶的反应（Cony et al.，2010）。乳清蛋白质经热处理或添加还原剂[如二硫苏糖醇（DTT）或半胱氨酸]变性后成为 TGase 的活性底物（Sharma et al.，2001；Tang and Ma，2007）。

通过蛋白质分子间的共价交联反应，采用 TGase 对乳清蛋白质的热稳定性进行了修饰（Agyare and Damodaran，2010；Wang et al.，2013b；Zhong et al.，2013）。TGase 处理的乳清蛋白质对 pH 非常敏感，这可能是由于赖氨酸残基正电荷的部分损失以及蛋白质表面的亲水-疏水平衡被破坏（Agyare and Damodaran，2010）。因此，TGase 处理过的乳清蛋白质不用于酸性饮料。

经 TGase 处理后，乳清蛋白质的凝胶温度和变性温度都有所提高（Chanasattru et al.，2007；Damodaran and Agyare，2013）。TGase 的处理时间、乳清蛋白质的预热方式、TGase 和乳清蛋白质的浓度都会影响在中性 pH 下加热蛋白的热稳定性（Chanasattru et al.，2007；Wang et al.，2013b；Zhong et al.，2013）。

TGase 通常用于交联蛋白质，导致聚合和凝胶化。TGase，尤其是微生物源 TGase（mTGase），是一种修饰一些食品的理化性质的有效工具，包括食品的流变性、力学性质、脱水收缩和乳化性能。在酸奶生产中 mTGase 的应用为终产品提供了一些很有前景的益处，包括提高凝固型酸奶的凝胶强度、减少凝固型酸奶的脱水收缩、改善搅拌型酸奶的黏度和乳化性状。

8.2.2 酶水解

作为一种商业化的产品，乳清蛋白质水解物（WPH）具有较高的营养价值、功能性和生物学特性。它们通常是由蛋白水解酶水解乳清蛋白质而产生并以水解度（%）为特征，通过比较断开的肽键数和肽键的总数来计算水解度（Pasupuleti et al.，2008）。六种不同的酶，即胰蛋白酶、胃蛋白酶、胰凝乳蛋白酶、木瓜蛋白酶、菠萝蛋白酶和链霉蛋白酶，可以用来水解乳清蛋白质。其中，胰蛋白酶、胃蛋白酶和胰凝乳蛋白酶主要用于生产 WPH。此外，可以通过预处理（均质条件）和水解过程[酶的种类、水解条件（酶底物比、温度、pH 和时间）]和环境条件来确定其功能性和生物活性（Jeewanthi et al.，2015）。

与乳清蛋白质相比，WPH 具有许多优点，包括热稳定性的提高、低致敏性、含有更多的生物活性肽以及为特殊饮食生产靶向肽。乳清蛋白质经酶水解后，其凝胶性、起泡性和乳化性均可得到改变（Foegeding et al.，2002；Wijayanti et al.，2014）。

天然乳清蛋白质由于其致密的球状结构，不易水解。在酶水解之前，预处理是非常重要的，这会导致先前被隐藏的更多的水解位点暴露出来，并且有更多的肽键可被攻击。因此，人们开发了不同的方法来提高乳清蛋白质的酶水解率，包括预热、亚硫酸分解、高压、介质极性改变和酯化反应等（Chobert et al.，2010；Kananen et al.，2000；Maynard et al.，1998；O'loughlin et al.，2012）。

高水解度的乳清蛋白质水解物扩大了乳清蛋白质在特殊食品中的应用范围，包括低过敏性婴儿配方食品、肠内配方食品和运动营养食品。

8.3 超声波处理

超声波是指频率超过人类听力极限（约 20 kHz）的声波（Gallego-Juárez et al.，2010）。根据频率范围，超声波可分为高强度（16~100 kHz，10~1000 W/cm^2）和低强度（从 100 kHz 到 1 MHz，强度低于 1 W/cm^2）两类。低强度超声波主要用于食品工业中的无损检测，以确保食品的高品质，包括硬度、成熟度、含糖量以及酸度（Soria and Villamiel，2010）。高强度（也称高功率或高能量）超声波具有破坏性，可用于材料改性，如香精提取、排气、消泡、乳化、微生物灭活和蛋白质改性（Awad et al.，2012）。

高强度超声波可以通过超声空化效应对食品的物理、化学或生化性质产生影响，通过空化气泡在液体中的崩塌，释放出大量高度局部化的能量（Soria and Villamiel，2010）。超声通过连续的压缩波和稀疏波传播。液体中的超声空化效应可以引起食品的化学和物理变化。这些气泡经历了形成、增长和内爆的过程。气泡内的高温是在空化气泡破裂时产生的,同时伴随着光(声致发光)的发射。液体中的超声空化效应也会产生冲击波、液体的湍流运动和自由基(Ashokkumar, 2011; Lauterborn et al., 2007)。

近年来,超声波技术被广泛应用于提高食品质量。HUS 能改变一些食品蛋白的物理和功能性质,包括表面疏水性、溶解性、流变性、乳化性和起泡性(Hu et al., 2013; O'Sullivan et al., 2015)。超声处理的这些变化与蛋白质的来源和加工参数密切相关。HUS 是一种有效的改变乳清蛋白质粒径的方法(Gordon and Pilosof, 2010; Shen et al., 2016a; Shen et al., 2016b)。使用 HUS 后,颗粒尺寸减小,这可能是由于与超声空化相关的流体剪切力极大地破坏了乳清蛋白质的共价和非共价相互作用。

HUS 也可提高乳清蛋白质的溶解性、起泡性和乳化性(Jambrak et al., 2008; Shen et al., 2016b)。此外,超声波可以降低乳清蛋白质溶液的黏度,改善乳清蛋白质溶液的热致凝胶性能(Chandrapala et al., 2011; Zisu et al., 2010)。另外,HUS 还可以提高葡萄糖酸内酯(GDL)诱导的 WPI 凝胶的持水能力、凝胶强度和硬度(Shen et al., 2017)。HUS 降低了预热 WPI 的粒径并且增加了其游离巯基的含量,这使得 GDL 诱导的 WPI 凝胶中形成了更多的二硫键,形成了致密均匀的凝胶网络,从而具有较高的乳清蛋白质浓度、凝胶强度和凝胶硬度。

HUS 对乳清蛋白质的改性受多种因素的影响,这些因素包括超声强度、超声时间、超声加工过程中是否进行温度控制以及乳清蛋白质中的其他组分。热处理和 HUS 的结合对乳清蛋白质的理化性质和乳化性能有很大的影响(Shen et al., 2016a)。热处理与 HUS 适当地结合使用可提高乳清蛋白质可溶性聚合物的热稳定性、流体特性和乳化性能。

HUS 被认为是一种环保、成本低、易于操作的技术。HUS 是一种有发展潜力的蛋白质改性技术。然而,在 HUS 使用过程中,为了获得更好的效果,最佳工艺条件包括频率、振幅、持续时间、温度、离子强度、pH 和蛋白质浓度以及来源等都需要进行标准化。此外,在商业中使用 HUS 时,还需考虑能源的输入问题。

8.4 高压处理

高压加工,又称高静压(HHP)加工或超高压加工,是一种以食品保鲜为主要目的的非热加工技术。在这种处理方法中,食品在短时间内暴露于压力(100~1000 MPa)下,以获得高品质的食品,包括微生物和酶的失活、营养素的保留以及根据需要改变食物属性。高压可以通过控制压力引起的食品成分(如蛋白质、多糖、淀粉等)的变化来改变食品的结构和质地。高压对蛋白质的改性是由疏水和静电相互作用的破坏引起的(Boonyaratanakornkit et al., 2002)。高压可以引起蛋白质结构的改变，导致变性/聚合，进一步形成凝胶或沉淀，这取决于蛋白质体系（蛋白质性质、浓度、溶液条件）和处理条件（压力水平、持续时间和温度）（Thom et al., 2006）。高压可以增加蛋白质的起泡性、乳化性和界面流变性（İbanoğlu et al., 2001; Lee et al., 2008; Yuan et al., 2013）。

在中性 pH 和室温下，乳清蛋白质组分对压力的敏感性为：β-乳球蛋白 B＞β-乳球蛋白 A＞牛血清白蛋白＞α-乳清白蛋白（Devi et al., 2013）。在一定条件下，高压可以诱导乳清蛋白质凝胶化。有很多因素影响高压导致的乳清蛋白质凝胶，包括压力水平、处理温度和时间、乳清蛋白质组成和浓度以及环境条件（pH 和离子强度）（He et al., 2010）。WPI 凝胶的硬度可以随着压力的升高、保压时间的延长或处理温度的升高而提高。WPI 在 25℃下形成凝胶的最低压力为 250 MPa（He and Kangcheng, 2009）。当 pH 为 6.8 时，形成凝胶的最低 WPI 浓度为 10%（Kanno et al., 1998）。此外增加乳清蛋白质的浓度也可以缩短凝胶时间。

超高压可导致乳清蛋白质的构象变化，导致疏水性增加和蛋白质聚合（Liu et al., 2005）。动态超高压增加了乳清蛋白质的表面疏水性。压力水平对乳清蛋白质的表面疏水性有很大的影响，随着压力水平从 0 MPa 到 300 MPa，疏水性逐渐增加。然而，动态超高压对乳清蛋白质的溶解度没有影响（Bouaouina et al., 2006）。

在乳化体系中，高压处理可以减少油滴并改变蛋白膜的组成。除了液滴大小外，高压处理还可以用来改变蛋白质的乳化性。由于压力诱导乳清蛋白质的聚合，高压处理降低了乳清蛋白质的乳化性（Galazka et al., 2000）。而高压处理可以提高 WPI-多糖混合物的乳化性（Yuan et al., 2013）。

在 pH 为 7.0 时，高压可以改善乳清蛋白质的起泡性，这可能是由于蛋白质的变性导致了一些分子间的相互作用，提高了蛋白质的吸附率。此外，压力水平和处理时间都会影响起泡性，并且处理时间似乎比 150～300 MPa 之间的压力水平影响更为显著（Bouaouina et al., 2006）。当压力超过 300 MPa 时，乳清蛋白质的起泡稳定性降低，这可能是由于蛋白质变性对泡沫稳定性带来的不利影响（İbanoğlu et al., 2001）。

HHP 作为一种创新的非热加工技术，最初用于食品保鲜。HHP 技术的应用引起了食品生产商的兴趣，不仅是因其在极低或中等温度下可高效杀灭微生物，而且在改善食品功能性和生物活性方面具有优越性。在过去的十几年中，高压技术在食品加工中的应用稳步增加。尽管高压对微生物的影响已经被研究了很多年，但更多的研究集中在它的潜在用途上，也就是生产新型的健康食品和有价值的材料。HHP 单独使用或与其他工艺流程一起使用，可以改变蛋白质的构象，定制乳清蛋白质使其发挥不同的功能。对于生产一系列新的低过敏性和高附加值乳清蛋白质基产品，HHP 技术被认为具有最高的潜力。

8.5 电脉冲处理

PEF 加工是一种非热加工食品保鲜技术。它利用短时高压电场使微生物和酶失活,从而延长食品的货架期。

对于蛋白质改性来说,蛋白质分子的去折叠是聚合的第一步。PEF 对蛋白质的影响可能涉及以下几个方面:①蛋白质分子的极化;②与四级结构相关的非共价连接的蛋白质亚基的解离;③蛋白质构象变化导致暴露先前隐藏的疏水性氨基酸或巯基;④极化结构往往通过静电引力相互吸引;⑤如果电脉冲的持续时间足够长,可能会通过疏水作用或共价键(即二硫键)形成蛋白质聚合物(Barsotti et al.,2001;Wei and Yang,2009;Zhao et al.,2012)。此外,PEF 下蛋白质的聚合与蛋白质浓度、pH、电场强度、电导率、暴露时间和温度有关(Perez and Pilosof,2004;Zhao and Yang,2012)。

用 PEF 处理牛奶时,酪蛋白微粒和乳清蛋白质的功能性质,如聚合性、疏水性、热稳定性、凝乳酶能力、凝胶速率、凝胶强度和乳化稳定性都会随 PEF 的强度、持续时间和温度而改变(Buckow et al.,2014)。

PEF 处理后的乳清蛋白质的结构取决于电场强度、脉冲数、蛋白质浓度以及处理过程中的温度。通过控制这些条件,PEF 处理可以用来生产具有部分功能特性的乳清蛋白质(Xiang et al.,2011;Xiang et al.,2008)。PEF 处理(30~35 kV/cm,19.2~211 μs,30~75℃)不影响 WPI 的任何理化或乳化性能。但在乳清蛋白质浓缩期间和喷雾干燥之前,相同条件下的 PEF 处理可破坏乳清蛋白质的沉淀和凝胶(Sui et al.,2011)。PEF 处理(12.5 kV/cm,10 次脉冲)β-LG(10%)改变了β-LG 浓缩物的天然结构,大大降低了β-LG 的热稳定性,提高了凝胶速率(Perez and Pilosof,2004)。然而,PEF 处理(21~36 kV/cm,低于 30℃)的β-LG(高达12%)并未导致蛋白质变性或聚合(Barsotti et al.,2001)。这些研究表明,在高场强下较长的持续时间比较短的持续时间更具有破坏性。

在连续模式下使用方波脉冲进行 PEF 处理(30~35 kV/cm),不会改变牛乳铁蛋白的理化性质(Qian et al.,2010)。然而,在 PEF 处理过程中,在较高温度(60~70℃)下,蛋白质聚合物表面疏水性增加,这在很大程度上是由相关的热效应引起的,也许部分是由 PEF 处理应力引起的。在 PEF 处理(40 kV/cm 和 1000 Hz)下,牛乳蛋白质(IgG、IgA 和 Lf)变性达到 70%,这可能是由 PEF 处理过程中的加工温度过高导致的(Alexander et al.,2014)。

PEF 的应用可能是热处理的一个很好的替代方法,特别是在蛋白质类食品中,因为蛋白质对 PEF 处理的敏感性不如热处理的巴氏杀菌。虽然 PEF 处理对蛋白质有轻微的改变，但是 PEF 处理可以改变蛋白质的二级和三级结构。如果 PEF 处理足够强，则可以在 PEF 处理后改变乳清蛋白质的功能特性。设备设计、处理条件和操作方式都会影响乳清蛋白质的功能特性。

8.6 辐 射 处 理

辐射处理是食品工业中保持产品新鲜度和质量的一种物理方法。除食品保鲜外，辐射技术作为一种非热、低成本的技术，可以应用于食品蛋白质的改性，特别是对蛋白膜和凝胶特性的改性。辐射后，羟基自由基和超氧阴离子自由基的产生会影响蛋白质的一级、二级和三级结构（Davies and Delsignore，1987）。辐射对蛋白质四级结构的影响取决于蛋白质浓度和存在的氧气（Lee et al.，2003）。辐射处理可以实现各种形式的改性，如蛋白质分子的交联（聚合）或降解（解聚），这与吸附的辐射剂量、辐射暴露时间、暴露条件和使用的食物蛋白质类型有关（Kuan et al.，2013）。

8.6.1 γ射线辐照

γ射线辐照在乳清蛋白质改性中的应用主要是基于对蛋白膜的处理（Cieśla et al.，2006；Cieśla et al.，2004）。一般而言，γ射线辐照是改善蛋白质可食用膜和涂层的阻隔性能和机械性能的有效方法（Cieśla et al.，2004）。辐照后蛋白质溶液的黏度因辐射交联而增加。用γ射线（32 kGy）处理乳清蛋白质溶液，可增加β链和β片的含量。γ辐射增加了乳清蛋白质凝胶和薄膜的断裂强度（Cieśla et al.，2006）。γ射线可以破坏蛋白质（BSA 和β-LG）结构以及多肽链的交联和聚合（Gaber，2005）。此外，辐照增加了乳清蛋白质颗粒的分子量，并降低了蛋白膜的透湿气性能（Ouattara et al.，2002）。

8.6.2 紫外线照射

在食品工业中，利用紫外线（UV）对蛋白质进行改性，以获得具有良好透气性和机械性能的蛋白膜。在碱性条件下，紫外辐射增加了乳清蛋白质的聚合，改变了膜的游离巯基、膜的色泽和溶解性。因此，紫外辐射和碱处理法的结合使用可以用于获得性能可适应多种食品包装应用的薄膜（Díaz et al.，2017）。紫外线照射也增加了乳清基薄膜的拉伸强度（Schmid et al.，2017）。但是，辐射处理对乳清蛋白质基薄膜的阻隔性能、断裂率和延伸率没有影响（Schmid et al.，2015；Zisu et al.，2010）。紫外线照射增加了乳清蛋白质膜的拉伸强度和褐变程度（取决于辐射时间）（Schmid et al.，2015）。在不同的热处理方式下紫外线辐射改变了薄膜的性质（Díaz et al.，2016）。紫外线辐射（270 nm 和 290 nm）可改变乳清蛋白质的结构，增加游离巯基的浓度（Kehoe et al.，2008；Permyakov et al.，2003；Vanhooren et al.，2002）。紫外线诱导的乳清蛋白质构象变化增加了乳清蛋白质对胃蛋白酶水解的敏感性（Kristo et al.，2012）。

此外，作为非电离辐射的微波对小麦面筋蛋白溶解度有负面影响（Yalcin et al.，2008）。微波和酶处理的应用降低了β-LG和WPI的免疫反应性（El Mecherfi et al.，2011）。然而，微波加热对乳清蛋白质功能性质的影响尚未得到充分的研究。

8.7 化学改性

食品蛋白质的化学改性主要涉及化学试剂的使用和美拉德反应。由于美拉德反应的自发性和无外来化学物质的加入，美拉德反应优于其他类型的食品蛋白质化学改性（如乙酰化、琥珀酰化、磷酸化、酯化、酰胺化和硫基化）。通过美拉德反应进行糖基化是改善功能的有效途径。因此，美拉德反应成为食品工业中改善蛋白质功能特性的一种很有前途的方法。

美拉德反应也称为非酶促褐变，是一种氨基和羰基间发生的天然的化学反应。在赖氨酸残基的ε-氨基和羰基之间主要形成共价键。美拉德反应在"湿"（溶液）和"干"条件下均可发生，在加热条件下可加速反应。由于蛋白质与糖链的共价连接，蛋白质的功能性质（如溶解度、热稳定性、乳化性和流变性）在接枝反应后会发生变化（Sun et al.，2011）。糖复合物的结构、理化性质和功能性质受水分活度（A_w）、pH、温度、时间、氨羰基反应比、食物蛋白质来源、还原糖（单糖、双糖、寡糖和多糖）性质的影响（Melton，2006）。

形成的糖复合物导致的乳清蛋白质的改性显示出更好的性能。乳清蛋白质的溶解性、热稳定性和乳化性可以通过与各种糖（主要是还原糖）糖基化而改变。WPC和羧甲基纤维素（CMC）在60℃干热条件下的美拉德反应产物，提高了乳清蛋白质的乳化性能。另外，美拉德反应产物WPC-CMC提高了乳液的热稳定性（Kika et al.，2007）。干热处理（两天培育期足够）使WPI与低甲氧基果胶（LMP）交联，可显著提高蛋白质的功能性质，特别是乳化活性和热稳定性。无论pH和离子强度如何，WPI-LMP结合物形成的稳定乳浊液具有更高的乳化特性和耐热稳定性（Setiowati et al.，2017）。在水溶液中WPI与葡聚糖（DX）反应得到的WPI-DX结合物的乳化能力、乳液稳定性、热稳定性均比WPI有所提高。而且在pH 3.2~7.5和离子强度0.05~0.2 mol/L范围内，结合物仍然是可溶的（Zhu et al.，2010）。理化性质/构象的变化（如变性温度、等电点、表面疏水性、巯基和独特的糖基化位点）导致分子间相互作用减少，最终导致WPI-DX结合物的溶解度和热稳定性的改善（Wang and Ismail，2012）。同样，通过美拉德反应生成WPI-甜菜果胶（SBP）

结合物，显著提高了 WPI 的热稳定性和乳化性能（Qi et al., 2017a，b）。WPI 与榴莲籽胶（DSG）通过部分结合过程提高了乳液的界面活性，是一种稳定水包油乳浊液的有效乳化剂（Tabatabaee and Mirhosseini, 2014）。此外，美拉德反应还可以改善乳清蛋白质的溶解性、热稳定性和乳化性，如α-LA-阿拉伯胶和β-LG-DX（de Oliveira et al., 2015；Jimenezcastano et al., 2005）。

在干燥状态下，α-LA 经美拉德反应与鼠李糖、岩藻糖糖基化，可提高其起泡稳定性（Haar et al., 2011）。通过美拉德反应，用葡萄糖或乳糖对β-LG 进行改性，形成具有更好起泡性能的糖化产物（Medrano et al., 2009）。美拉德反应显著改善了 WPI-DX 凝胶的力学特性（Spotti et al., 2013；Sun et al., 2011）。WPI/DX 结合凝胶比 WPI 天然体系或 WPI/DX 混合体系形成的凝胶弱得多。与 WPI 相比，WPI/DX 混合体系的凝胶化时间和温度都有所降低（Spotti et al., 2014）。此外，美拉德反应产物还具有很高的抗氧化活性，这取决于改性所用的糖的类型（Wang et al., 2013a）。美拉德反应可以提高 WPI 的抗氧化能力（Liu et al., 2014a，b）。

8.8 总　　结

乳清蛋白质的几种改性方法在食品和生物技术领域有着广泛的应用，特别是由于它们能够改变乳清蛋白质的理化性质和功能特性。本章讨论了不同的乳清蛋白质改性方法及其对乳清蛋白质理化和功能特性的影响。这些改性方法包括传统加热、酶处理、化学改性和一些新型的非热食品加工技术。利用这些方法来改变乳清蛋白质的某些特性以扩大其在食品和非食品领域中的应用，已经取得了相当大的进展。然而，为了提高乳清蛋白质在食品工业中的进一步应用，还需要深入研究。

参 考 文 献

Agyare, K.K. and Damodaran, S. (2010). pH-stability and thermal properties of microbial transglutaminase-treated whey protein isolate. *Journal of Agricultural and Food Chemistry* **58** (3): 1946-1953.

Alexander, M., Stefan, T., Claudia, S. et al. (2014). Pulsed electric field treatment process and dasry product comprising bioactive molecules obtainable by the process. US Patent 20,140,154,371A1, filed 6 July 2012 and issued 5 June 2014.

Ashokkumar, M. (2011). The characterization of acoustic cavitation bubbles-an overview. *Ultrasonics Sonochemistry* **18** (4): 864-872.

Awad, T.S., Moharram, H.A., Shaltout, O.E. et al. (2012). Applications of ultrasound in analysis, processing and quality control of food: a review. *Food Research International* **48** (2): 410-427.

Barsotti, L., Dumay, E., Mu, T.H. et al. (2001). Effects of high voltage electric pulses on protein-based food constituents and structures. *Trends in Food Science & Technology* **12** (3-4): 136-144.

Boonyaratanakornkit, B.B., Park, C.B., and Clark, D.S. (2002). Pressure effects on intra- and intermolecular interactions within proteins. *Biochimica et Biophysica Acta (BBA) - Protein Structure and Molecular Enzymology* **1595** (1-2): 235-249.

Bouaouina, H., Desrumaux, A., Loisel, C. et al. (2006). Functional properties of whey proteins as affected by dynamic high-pressure treatment. *International Dairy Journal* **16** (4): 275-284.

Buckow, R., Chandry, P.S., Ng, S.Y. et al. (2014). Opportunities and challenges in pulsed electric field processing of dairy products. *International Dairy Journal* **34** (2): 199-212.

Chanasattru, W., Decker, E.A., and McClements, D.J. (2007). Modulation of thermal stability and heat-induced gelation of β-lactoglobulin by high glycerol and sorbitol levels. *Food Chemistry* **103** (2): 512-520.

Chandrapala, J., Zisu, B., Palmer, M. et al. (2011). Effects of ultrasound on the thermal and structural characteristics of proteins in reconstituted whey protein concentrate. *Ultrasonics Sonochemistry* **18** (5): 951-957.

Chobert, J.M., Briand, L., Dufour, E. et al. (2010). How to increase β-lactoglobulin susceptibility to peptic hydrolysis. *Journal of Food Biochemistry* **20** (4): 439-462.

Cieśla, K., Salmieri, S., and Lacroix, M. (2006). Gamma-irradiation influence on the structure and properties of calcium caseinate-whey protein isolate based films. Part 2. Influence of polysaccharide addition and radiation treatment on the structure and functional properties of the films. *Journal of Agricultural and Food Chemistry* **54** (23): 8899-8908.

Cieśla, K., Salmieri, S., Lacroix, M. et al. (2004). Gamma irradiation influence on physical properties of milk proteins. *Radiation Physics and Chemistry* **71** (1-2): 95-99.

Cony, G., Pedrolm, B., and Marildet, B.L. (2010). Effect of thermal treatment on whey protein polymerization by transglutaminase: implications for functionality in processed dairy foods. *LWT-Food Science and Technology* **43** (2): 214-219.

Díaz, O., Candia, D., and Cobos, Á. (2016). Effects of ultraviolet radiation on properties of films from whey protein concentrate treated before or after film formation. *Food Hydrocolloids* **55**: 189-199.

Díaz, O., Candia, D., and Cobos, Á. (2017). Whey protein film properties as affected by ultraviolet treatment under alkaline conditions. *International Dairy Journal* **73**: 84-91.

Damodaran, S. and Agyare, K.K. (2013). Effect of microbial transglutaminase treatment on thermal stability and pH-solubility of heat-shocked whey protein isolate. *Food Hydrocolloids* **30** (1): 12-18.

Davies, K.J.A. and Delsignore, M.E. (1987). Protein damage and degradation by oxygen radicals. III. Modification of secondary and tertiary structure. *Journal of Biological Chemistry* **262** (20): 9908-9913.

de Oliveira, F.C., Js, D.R.C., de Oliveira, E.B. et al. (2015). Acacia gum as modifier of thermal stability, solubility and emulsifying properties of α-lactalbumin. *Carbohydrate Polymers* **119**: 210-218.

Devi, A.F., Buckow, R., Hemar, Y. et al. (2013). Structuring dairy systems through high pressure processing. *Journal of Food Engineering* **114** (1): 106-122.

Eallen, F. and Jackp, D. (2011). Food protein functionality: a comprehensive approach. *Food Hydrocolloids* **25** (8): 1853-1864.

El Mecherfi, K.E., Saidi, D., Kheroua, O. et al. (2011). Combined microwave and enzymatic treatments for β-lactoglobulin and bovine whey proteins and their effect on the IgE immunoreactivity. *European Food Research & Technology* **233** (5): 859-867.

Foegeding, E.A., Davis, J.P., Doucet, D. et al. (2002). Advances in modifying and understanding whey protein functionality. *Trends in Food Science & Technology* **13** (5): 151-159.

Gaber, M.H. (2005). Effect of gamma-irradiation on the molecular properties of bovine serum albumin. *Journal of Bioscience and Bioengineering* **100** (2): 203-206.

Galazka, V.B., Dickinson, E., and Ledward, D.A. (2000). Influence of high pressure processing on protein solutions and emulsions. *Current Opinion in Colloid & Interface Science* **5** (3-4): 182-187.

Gallego-Juárez, J.A., Rodriguez, G., Acosta, V. et al. (2010). Power ultrasonic transducers with extensive radiators for industrial processing. *Ultrasonics Sonochemistry* **17** (6): 953-964.

Gordon, L. and Pilosof, A.M.R. (2010). Application of high-intensity ultrasounds to control the size of whey proteins particles. *Food Biophysics* **5** (3): 203-210.

Guo, M. and Wang, G. (2016). Whey protein polymerisation and its applications in environmentally safe adhesives. *International Journal of Dairy Technology* **69** (4): 481-488.

Haar, R.T., Westphal, Y., Wierenga, P.A. et al. (2011). Cross-linking behavior and foaming properties of bovine α-lactalbumin after glycation with various saccharides. *Journal of Agricultural and Food Chemistry* **59** (23): 12460-12466.

He, J.S., Azuma, N., and Yang, H. (2010). Effects of pH and ionic strength on the rheology and microstructure of a pressure-induced whey protein gel. *International Dairy Journal* **20** (2): 89-95.

He, J.S. and Kangcheng, R. (2009). Kinetics of phase separation during pressure-induced gelation of a whey protein isolate. *Food Hydrocolloids* **23** (7): 1729-1733.

Hu, H., Wu, J., Li-Chan, E.C.Y. et al. (2013). Effects of ultrasound on structural and physical properties of soy protein isolate (SPI) dispersions. *Food Hydrocolloids* **30** (2): 647-655.

Hussain, R., Gaiani, C., Jeandel, C. et al. (2012). Combined effect of heat treatment and ionic strength on the functionality of whey proteins. *Journal of Dairy Science* **95** (11): 6260-6273.

İbanoğlu, E., Karataş, Ş., İbanog˘lu, E. et al. (2001). High pressure effect on foaming behaviour of whey protein isolate. *Journal of Food Engineering* **47** (1): 31-36.

Jambrak, A.R., Mason, T.J., Lelas, V. et al. (2008). Effect of ultrasound treatment on solubility and foaming properties of whey protein suspensions. *Journal of Food Engineering* **86** (2): 281-287.

Jaros, D., Partschefeld, C., Henle, T. et al. (2006). Transglutaminase in dairy products: chemistry, physics, applications. *Journal of Texture Studies* **37** (2): 113-155.

Jeewanthi, R.K.C., Lee, N.K., and Paik, H.D. (2015). Improved functional characteristics of whey protein hydrolysates in food iIndustry. *Korean Journal for Food Science of Animal Resources* **35** (3): 350-359.

Jimenezcastano, L., Lopezfandino, R., Olano, A. et al. (2005). Study on the beta-lactoglobulin glycosylation with dextran: effect on solubility and heat stability. *Food Chemistry* **93** (4): 689-695.

Kananen, A., Savolainen, J., Mäkinen, J. et al. (2000). Influence of chemical modification of whey protein conformation on hydrolysis with pepsin and trypsin. *International Dairy Journal* **10** (10): 691-697.

Kanno, C., Mu, T.H., Hagiwara, T. et al. (1998). Gel formation from industrial milk whey proteins under hydrostatic pressure: effect of hydrostatic pressure and protein concentration. *Journal of Agricultural and Food Chemistry* **46** (2): 417-424.

Kehoe, J.J., Remondetto, G.E., Subirade, M. et al. (2008). Tryptophan-mediated denaturation of β-lactoglobulin a by UV irradiation. *Journal of Agricultural and Food Chemistry* **56** (12): 4720-4725.

Kika, K., Korlos, F., and Kiosseoglou, V. (2007). Improvement, by dry-heating, of the emulsionstabilizing properties of a whey protein concentrate obtained through carboxymethylcellulose complexation. *Food Chemistry* **104** (3): 1153-1159.

Kim, D.A., Cornec, M., and Narsimhan, G. (2005). Effect of thermal treatment on interfacial properties of β-lactoglobulin. *Journal of Colloid and Interface Science* **285** (1): 100-109.

Kristo, E., Hazizaj, A., and Corredig, M. (2012). Structural changes imposed on whey proteins by UV irradiation in a continuous UV light reactor. *Journal of Agricultural and Food Chemistry* **60** (24): 6204-6209.

Kuan, Y.H., Bhat, R., Patras, A. et al. (2013). Radiation processing of food proteins-a review on the recent developments. *Trends in Food Science & Technology* **30** (2): 105-120.

Kuraishi, C., Yamazaki, K., and Susa, Y. (2001). Transglutaminase: its utilization in the food industry. *Food Reviews International* **17** (2): 221-246.

Lauterborn, W., Kurz, T., Geisler, R. et al. (2007). Acoustic cavitation, bubble dynamics and sonoluminescence. *Ultrasonics Sonochemistry* **14** (4): 484-491.

Lee, S., Lee, S., and Song, K.B. (2003). Effect of gamma-irradiation on the physicochemical properties of porcine and bovine blood plasma proteins. *Food Chemistry* **82** (4): 521-526.

Lee, S.H., Subirade, M., and Paquin, P. (2008). Effects of ultra-high pressure homogenization on the emulsifying properties of whey protein isolates under various pH. *Food Science and Biotechnology* **17** (2): 324-329.

Leila, M., Mahdi, K., and Mohammad, S. (2009). Effects of succinylation and deamidation on functional properties of oat protein isolate. *Food Chemistry* **114** (1): 127-131.

Liu, Q., Kong, B., Han, J. et al. (2014a). Structure and antioxidant activity of whey protein isolate conjugated with glucose via the Maillard reaction under dry-heating conditions. *Food Structure* **1** (2): 145-154.

Liu, Q., Li, J., Kong, B. et al. (2014b). Physicochemical and antioxidant properties of Maillard reaction products formed by heating whey protein isolate and reducing sugars. *International Journal of Dairy Technology* **67** (2): 220-228.

Liu, X., Powers, J.R., and Swanson, B.G. (2005). Modification of whey protein concentrate hydrophobicity by high hydrostatic pressure. *Innovative Food Science & Emerging Technologies* **6** (3): 310-317.

Maynard, F., Weingand, A., Hau, J. et al. (1998). Effect of high-pressure treatment on the tryptic hydrolysis of bovine beta-lactoglobulin AB. *International Dairy Journal* **8** (2): 125-133.

Medrano, A., Abirached, C., Panizzolo, L. et al. (2009). The effect of glycation on foam and structural properties of β-lactoglobulin. *Food Chemistry* **113** (1): 127-133.

Melton, L.D. (2006). Creating proteins with novel functionality via the Maillard reaction: a review. *Critical Reviews in Food Science & Nutrition* **46** (4): 337-350.

Mirmoghtadaie, L., Shojaee, A.S., and Hosseini, S.M. (2016). Recent approaches in physical modification of protein functionality. *Food Chemistry* **199**: 619-627.

Nicolai, T., Britten, M., and Schmitt, C. (2011). β-Lactoglobulin and WPI aggregates: formation, structure and applications. *Food Hydrocolloids* **25** (8): 1945-1962.

Nicolai, T. and Durand, D. (2013). Controlled food protein aggregation for new functionality. *Current Opinion in Colloid & Interface Science* **18** (4): 249-256.

O'Loughlin, I.B., Murray, B.A., Kelly, P.M. et al. (2012). Enzymatic hydrolysis of heat-induced aggregates of whey protein isolate. *Journal of Agricultural and Food Chemistry* **60** (19): 4895-4904.

O'Sullivan, J., Murray, B., Flynn, C. et al. (2015). The effect of ultrasound treatment on the structural, physical and emulsifying properties of animal and vegetable proteins. *Food Hydrocolloids* **53**: 141-154.

Ouattara, B., Canh, L.T., Vachon, C. et al. (2002). Use of γ-irradiation cross-linking to improve the water vapor permeability and the chemical stability of milk protein films. *Radiation Physics and Chemistry* **63** (3): 821-825.

Pasupuleti, V.K., Holmes, C., and Demain, A.L. (2008). Applications of protein hydrolysates in biotechnology. In: *Protein Hydrolysates in Biotechnology* (ed. V.K. Pasupuleti and A.L. Demain), 1-9. Dordrecht: Springer.

Perez, O.E. and Pilosof, A.M.R. (2004). Pulsed electric fields effects on the molecular structure and gelation of β-lactoglobulin concentrate and egg white. *Food Research International* **37** (1): 102-110.

Permyakov, E.A., Permyakov, S.E., Deikus, G.Y. et al. (2003). Ultraviolet illumination-induced reduction of alpha-lactalbumin disulfide bridges. *Proteins-structure Function & Bioinformatics* **51** (4): 498-503.

Qi, P.X., Xiao, Y., and Wickham, E.D. (2017a). Changes in physical, chemical and functional properties of whey protein isolate (WPI) and sugar beet pectin (SBP) conjugates formed by controlled dry-heating. *Food Hydrocolloids* **69**: 86-96.

Qi, P.X., Xiao, Y., and Wickham, E.D. (2017b). Stabilization of whey protein isolate (WPI) through interactions with sugar beet pectin (SBP) induced by controlled dry-heating. *Food Hydrocolloids* **67**: 1-13.

Qian, S., Roginski, H., Williams, R.P.W. et al. (2010). Effect of pulsed electric field and thermal treatment on the physicochemical properties of lactoferrin with different iron saturation levels.

International Dairy Journal **20** (10): 707-714.

Ryan, K.N. and Foegeding, E.A. (2015). Formation of soluble whey protein aggregates and their stability in beverages. *Food Hydrocolloids* **43** (43): 265-274.

Ryan, K.N., Zhong, Q., and Foegeding, E.A. (2013). Use of whey protein soluble aggregates for thermal stability-a hypothesis paper. *Journal of Food Science* **78** (8): R1105-R1115.

Schmid, M., Held, J., Hammann, F. et al. (2015). Effect of UV-radiation on the packaging-related properties of whey protein isolate based films and coatings. *Packaging Technology and Science* **28** (10): 883-899.

Schmid, M., Prinz, T.K., Müller, K. et al. (2017). UV radiation induced cross-linking of whey protein isolate-based films. *International Journal of Ploymer Science* https://doi.org/10.1155/2017/1846031.

Schmitt, C., Bovay, C., Rouvet, M. et al. (2007). Whey protein soluble aggregates from heating with NaCl: physicochemical, interfacial, and foaming properties. *Langmuir the Acs Journal of Surfaces & Colloids* **23** (8): 4155-4166.

Setiowati, A.D., Saeedi, S., Wijaya, W. et al. (2017). Improved heat stability of whey protein isolate stabilized emulsions via dry heat treatment of WPI and low methoxyl pectin: effect of pectin concentration, pH, and ionic strength. *Food Hydrocolloids* **63**: 716-726.

Sharma, R., Lorenzen, P.C., and Qvist, K.B. (2001). Influence of transglutaminase treatment of skim milk on the formation of epsilon-(gamma-glutamyl)lysine and the susceptibility of individual proteins towards crosslinking. *International Dairy Journal* **11** (10): 785-793.

Shen, X., Fang, T., Gao, F. et al. (2016a). Effects of ultrasound treatment on physicochemical and emulsifying properties of whey proteins pre- and post-thermal aggregation. *Food Hydrocolloids* **63**: 668-676.

Shen, X., Shao, S., and Guo, M. (2016b). Ultrasound-induced changes in physical and functional properties of whey proteins. *International Journal of Food Science and Technology* **52** (2): 381-388.

Shen, X., Zhao, C., and Guo, M. (2017). Effects of high intensity ultrasound on acid-induced gelation properties of whey protein gel. *Ultrasonics Sonochemistry* **39**: 810-815.

Smithers, G.W. (2015). Whey-ing up the options-yesterday, today and tomorrow. *International Dairy Journal* **48**: 2-14.

Soria, A.C. and Villamiel, M. (2010). Effect of ultrasound on the technological properties and bioactivity of food: a review. *Trends in Food Science & Technology* **21** (7): 323-331.

Spotti, M.J., Martinez, M.J., Pilosof, A.M.R. et al. (2014). Rheological properties of whey protein and dextran conjugates at different reaction times. *Food Hydrocolloids* **38** (38): 76-84.

Spotti, M.J., Perduca, M.J., Piagentini, A. et al. (2013). Gel mechanical properties of milk whey protein-dextran conjugates obtained by Maillard reaction. *Food Hydrocolloids* **31** (1): 26-32.

Sui, Q., Roginski, H., Rpw, W. et al. (2011). Effect of pulsed electric field and thermal treatment on the physicochemical and functional properties of whey protein isolate. *International Dairy Journal* **21** (4): 206-213.

Sun, W.W., Yu, S.J., Yang, X.Q. et al. (2011). Study on the rheological properties of heat-induced whey protein isolate-dextran conjugate gel. *Food Research International* **44** (10): 3259-3263.

Tabatabaee, A.B. and Mirhosseini, H. (2014). Stabilization of water in oil in water (W/O/W) emulsion using whey protein isolate-conjugated durian seed gum: enhancement of interfacial activity through conjugation process. *Colloids & Surfaces B Biointerfaces* **113** (1): 107-114.

Tang, C.H. and Ma, C.Y. (2007). Modulation of the thermal stability of β-lactoglobulin by transglutaminase treatment. *European Food Research and Technology* **225** (5-6): 649-652.

Thom, H., Marya, S., Vivekk, U. et al. (2006). High-pressure-induced changes in bovine milk: a review. *Biochimica Et Biophysica Acta Proteins & Proteomics* **1764** (3): 593-598.

Vanhooren, A., Devreese, B., Vanhee, K. et al. (2002). Photoexcitation of tryptophan groups induces reduction of two disulfide bonds in goat alpha-lactalbumin. *Biochemistry* **41** (36): 11035-11043.

Vardhanabhuti, B., Foegeding, E.A., Mcguffy, M.K. et al. (2001). Gelation properties of dispersions containing polymerized and native whey protein isolate. *Food Hydrocolloids* **15** (2): 165-175.

Wang, Q. and Ismail, B. (2012). Effect of Maillard-induced glycosylation on the nutritional quality, solubility, thermal stability and molecular configuration of whey proteinv. *International Dairy Journal* **27** (1-2): 112-122.

Wang, W.Q., Bao, Y.H., and Chen, Y. (2013a). Characteristics and antioxidant activity of water-soluble Maillard reaction products from interactions in a whey protein isolate and sugars system. *Food Chemistry* **139** (1-4): 355-361.

Wang, W., Zhong, Q., and Hu, Z. (2013b). Nanoscale understanding of thermal aggregation of whey protein pretreated by transglutaminase. *Journal of Agricultural and Food Chemistry* **61** (2): 435-446.

Wei, Z. and Yang, R. (2009). Effect of high-intensity pulsed electric fields on the activity, conformation and self-aggregation of pepsin. *Food Chemistry* **114** (3): 777-781.

Wijayanti, H.B., Bansal, N., and Deeth, H.C. (2014). Stability of whey proteins during thermal processing: a review. *Comprehensive Reviews in Food Science & Food Safety* **13** (6): 1235-1251.

Xiang, B.Y., Ngadi, M.O., and Simpson, M.V. (2011). Pulsed electric field-induced structural modification of whey protein isolate. *Food and Bioprocess Technology* **4** (8): 1341-1348.

Xiang, B.Y., Ngadi, M.O., Simpson, M.V. et al. (2008). Effect of pulsed electric field on structural modification and thermal properties of whey protein isolate. Presented in CSBE/SCGAB 2008 Annual Conference, North Vancouver, British Columbia (13-6 July 2008). Canada: The Canadian Society for Bioengineering.

Yalcin, E., Sakiyan, O., Sumnu, G. et al. (2008). Functional properties of microwave-treated wheat gluten. *European Food Research and Technology* **227** (5): 1411.

Yuan, F., Xu, D., Qi, X. et al. (2013). Impact of high hydrostatic pressure on the emulsifying properties of whey protein isolate-chitosan mixtures. *Food & Bioprocess Technology* **6** (4): 1024-1031.

Zhao, W. and Yang, R. (2012). Pulsed electric field induced aggregation of food proteins: ovalbumin and bovine serum albumin. *Food & Bioprocess Technology* **5** (5): 1706-1714.

Zhao, W., Yang, R., and Zhang, H.Q. (2012). Recent advances in the action of pulsed electric fields on enzymes and food component proteins. *Trends in Food Science & Technology* **27** (2): 83-96.

Zhong, Q., Wang, W., Hu, Z. et al. (2013). Sequential preheating and transglutaminase pretreatments improve stability ofwhey protein isolate at pH 7.0 during thermal sterilization. *Food Hydrocolloids* **31** (2): 306-316.

Zhu, D., Damodaran, S., and Lucey, J.A. (2010). Physicochemical and emulsifying properties of whey protein isolate (WPI)-dextran conjugates produced in aqueous solution. *Journal of Agricultural and Food Chemistry* **58** (5): 2988-2994.

Zisu, B., Bhaskaracharya, R., Kentish, S., and Ashokkumar, M. (2010). Ultrasonic processing of dairy systems in large scale reactors. *Ultrasonics Sonochemistry* **17** (6): 1075.

第 9 章 乳清蛋白质在非食品领域的应用

Mingruo Guo[1,2], Wenbo Wang[3], Zhenhua Gao[4],
Guorong Wang[1], and Liang Li[2]

1. Department of Nutrition and Food Science, University of Vermont, Burlington, USA
2. College of Food Science, Northeast Agriculture University, Harbin, People's Republic of China
3. Academy of Forest Inventory and Planning, Beijing, People's Republic of China
4. College of Materials Science and Engineering, Jilin University, Northeast Forest University, Harbin, People's Republic of China

9.1 胶水黏合理论

乳清作为一种干酪加工的副产品，曾经被当污水直接排放。如今随着乳清综合利用技术的提高，乳清蛋白质早已不再是一种蛋白废料了。相反，除了食品领域外，其在很多非食品的环保生物原料中也得到了很多应用，如乳清蛋白质胶水，包括木清漆、木胶、办公胶水以及手术胶水等。乳清蛋白质含有大量的功能基团（主要包括氨基、羟基和羧基），这些基团可以经过改性和交联，使其分子量增加并形成稳定的三维网状结构，产生足够的机械强度、黏合力、防水性。另外，与其他蛋白质（如大豆蛋白）相比，乳清蛋白质还具有两个明显的优势，就是热凝胶性和易水溶性。当乳清蛋白质溶液加热到 60℃时，就会通过分子内或分子间的巯基-二硫键交换或者巯基-巯基氧化反应交联成三维网状结构（Monahan et al., 1995；Dunkerly and Zadow, 1984）。这种热凝胶的形成赋予了乳清蛋白质作为胶水原料所需的黏合力和耐久性。但相比其他合成胶水原料相比，乳清蛋白质本身的黏合强度和磨光性相对弱一些，因此需要进行相应的蛋白改性。与此同时，乳清蛋白质在水中的溶解度可达 40%以上，使其不仅可以生产环保型产品，也可以通过其他技术手段进行化学改性、交联、混合。

涂漆是用于涂布到物体表面起到装饰、防水、防腐、增加耐用性等功能的材料。胶水是用于涂抹到物体表面从而将两个物体黏合并防止其分离的黏合剂（Guo and Wang, 2016a, b）。无论是涂漆还是胶水，产品的黏合力至关重要，因此理解涂料和胶水成分的黏合原理非常重要。相关的黏合理论包括吸附理论（adsorption theory)、机械互锁理论（mechanical interlocking）、化学键合理论（chemical bonding）、电子理论（electronic theory）、扩散理论（diffusion theory）、协作理论（coordination theory）、边界层理论（boundary layers' theory）等。本章将介绍前三种与乳清蛋白质胶水相关的理论。

9.1.1 吸附理论

吸附理论是指当胶水和黏合物表面紧密接触时，胶水的原子或分子与黏合物表面的原子或分子相互作用来获得黏合力。分子间的相互作用分为两类。第一类包括离子键、共价键、金属键这类化学吸附形成的永久结合力；第二类是范德瓦耳斯力、疏水作用力、氢键这类比较弱的物理吸附力。虽然物理吸附力比较弱，但是乳清蛋白质具有的大量极性基团可以与黏合面的极性基团（如木板中纤维分子中的氢键）形成氢键。物理吸附力（如范德瓦耳斯力和氢键）容易被水分子破坏，因此在干燥环境下黏合力比较强，但在潮湿环境下强度减弱，抗水性较差。

9.1.2 机械互锁理论

机械连锁理论是 MacBain 和 Hopkins 在 1925 年提出的。胶水或涂料的分子可以渗入到黏合物的空穴、小孔、凹凸中形成机械互锁从而产生黏合力。对于多孔材料，如木、纸、纺织品、陶器等材料，液体胶水或涂料会充分渗入到黏合表面的微小孔穴中，在胶水固化后形成类似钉子和铆钉的效果，产生较强的黏合力。根据这个理论，黏合面越粗糙，液体胶水渗入黏合物表面越深，胶水和黏合物的互锁力就越强。机械互锁理论也被广泛用来解释各种胶水黏合现象。

9.1.3 化学键合理论

如果胶水或涂料分子可以与黏合物材料的分子形成化学结合，无疑会大大增强黏合力和耐受力。因为化学键的能量级在 100～1000 kJ/mol 之间，而范德瓦耳斯力和氢键这类物理吸附力的能量小于 50 kJ/mol。但不是所有的胶水或涂料都能和黏合物形成化学键，只有在这两种材料都具有可互相反应的活性基团时，化学键才能形成。一个典型的例子就是木质材料与多异氰酸酯-聚亚甲基二苯基甲烷二异氰酸酯（p-MDI）胶水的结合。p-MDI 的异氰基与木纤维分子的羟基反应形成氨基甲酸酯键。与其他木用胶水相比（如尿素-甲醛胶水），其具有结合力强、胶水用量少、抗水性好的特点。如果化学键形成得足够多，就可以形成非常强的结合力，从而能抵抗更极端的环境或气候条件。在乳清蛋白质中引入化学键将是提高其胶水或涂料结合力和耐受力的有效途径之一。

9.2 木 清 漆

很多原料都可以用于各类木制品（如家具、玩具、地板、内部装饰）的涂层或磨光。这类涂料通常都是由石化原料合成的，如丙烯酸酯或氨基甲酸酯涂层，一般含有一定的有机挥发物（VOCs）。把涂料涂施到基板表面，溶剂蒸发后就形成一层具有保护或装饰作用的涂层。有些用于溶解、分散或者稀释有机树脂的溶剂是有毒性的，可能会对人体特别是儿童造成潜在的健康威胁。

乳清蛋白质是一种天然的高分子材料混合物。以乳清蛋白质的主要成分β-乳球蛋白为例，β-乳球蛋白的分子质量为 18.3 kDa，占乳清蛋白质总质量的一半左右。乳清蛋白质含有大量的极性基团，如氨基、羟基、羧基等。乳清蛋白质的水溶液在基板上形成薄层，当水分蒸发时，乳清蛋白质分子与基板材料分子的极性基团形成化学吸附和机械互锁，从而形成牢固的涂层。早在 2000 年，美国佛蒙特大学就已经成功研发了一种基于乳清蛋白质的木清漆并且实现了商业化应用。

跟其他商业化清漆一样，乳清蛋白质木清漆可以有不同的形态，如液体、乳化液、悬浮液等。这取决于产品组分在水中的不同溶解性和分散性。除了乳清蛋白质，还可以包括其他多糖类（纤维素、淀粉）和蛋白类（胶原蛋白、明胶、大豆蛋白、面筋蛋白、玉米蛋白、酪蛋白、酪朊酸钠、蛋清原料等）。乳清蛋白质可以是 WPI 或者 WPC，因为 WPC 中含有一定量脂肪和其他杂质，一定程度上会影响清漆的质量。相比之下，WPI 由于蛋白含量更纯、杂质少，因此做成的木清漆质量最好，具有良好的耐久性和透明度，但价格也相对比较高。

用于做木清漆的乳清蛋白质溶液的黏度可达 1000 mPa·s 以上。乳清蛋白质的黏度越高，所制成的清漆的涂抹性越好，涂层的厚度越合适，也因此对木板起到越好的保护作用。通常有两种方法可以增加木清漆的黏度。第一种方法是可在木清漆产品中添加一些 WPI 粉；第二种方法是可以对 WPI 溶液进行热处理（如 20%的乳清蛋白质溶液在 63℃加热 35~60 min，或者 10%的乳清蛋白质溶液在 90℃加热 30 min）。热处理可以使乳清蛋白质变性，形成蛋白分子交联。这种交联反应可以增加产品的黏度和耐久性。

乳清蛋白质同时具有亲水性和疏水性，因此乳清蛋白质粉或者溶液可以很好地与合成树脂（丙烯酸树脂、氨基甲酸酯等）混合。这种混合清漆同时具有合成漆的质量，也可以有效减少有机挥发成分。通过乳清蛋白质替代，合成树脂的用量可以从 70%降低到 30%。当乳清蛋白质与合成树脂的比例在 1∶2~1∶4 之间时，产品的硬度、色泽、黏性、涂抹性、防水性以及成本最佳。某些商业合成树脂的有机挥发成分达 250~450 g/L，使用乳清蛋白质可以有效降低有机溶剂的用量,从而使有机挥发溶剂降至 60～100 g/L。乳清蛋白质木清漆具有保护、防水、防刮、环保等特点,在木制品上涂层无色、透明、光滑、有亮光,厚度大于 5 μm,通常在 10～40 μm 之间。

9.3 木 胶 水

胶水的应用历史悠久,可追溯到古代埃及和中国,当时人们用胶水来粘纸莎草芦苇和制作木质家具。早期的胶水材料都是天然的碳水化合物或蛋白质,如淀粉、动物血以及从动物骨和皮中提取的胶原蛋白。后来牛乳蛋白和鱼类提取物也被作为胶水的原材料。再后来其他生物材料,如纤维素、大豆蛋白、木质素、单宁、淀粉、树皮也被用于制作胶水来黏合纸、木等材料(USDA Forest Products Laboratory,1987;Gao et al.,2007)。现代世界上的胶水原料主要是合成的聚合物材料(Gao et al.,2007;Pocius,2012)。

在合成木胶水材料中,甲醛类胶水是最重要的一种(Stoeckel et al.,2013)。甲醛类胶水含甲醛以及与甲醛反应的成分,包括尿素(UF)、三聚氰胺(MF)、苯酚(PF)、间苯二酚(RF)等及其混合胶[如三聚氰胺改性脲醛树脂(MUF)和苯酚-间苯二酚-甲醛(PRF)胶黏剂](Meyer and Hermans,1986)。除了甲醛胶水之外,异氰酸酯类胶水也有广泛的应用,这类胶水包括聚亚甲基二苯基甲烷二异氰酸酯(p-MDI)、乳化型异氰酸酯聚合物(EPI)、反应型聚氨酯(PUR)胶和聚乙酸乙烯酯(PVAc)胶,在木材和家具行业中普遍应用。

大部分商业胶水中的合成树脂可能对人体健康有害。首先是在产品的生产、运输和使用过程中产生挥发性有机物,如甲醛(Meyer and Hermans,1986;Henderson,1979)。甲醛挥发会刺激眼睛和喉咙,造成呼吸道不适,也是潜在的致癌物(Perera and Petito,1982;Swenberg et al.,1980)。其次是苯酚、异氰酸酯单体、乙烯单体或者其他有机溶剂等有机挥发物。随着对甲醛等有机挥发物的健康危害关注越来越多,人们对安全环保型胶水的需求也越来越多。再次是这些合成树脂的生产离不开石化原料,石化原料作为一种不可再生资源,储量有限,同时造成很多环境污染。另外,合成树脂不能被降解,对环境的污染不容忽视。由于合成树脂对人体的健康危害和对石化工业的重度依赖,从 1980 年开始,越来越多的研究开始寻找可再生原料来替代石化原料制作木胶水(Hemingway et al.,1989)。

乳清蛋白质是分子量较小的球蛋白混合物(与酪蛋白分子量比),紧凑折叠的球蛋白结构其实是不适宜用于胶水原料的。但是乳清蛋白质的以下几个特点又决定其具有作为一种胶水原料的可能性。首先,乳清蛋白质具有优良的溶解性,在

45~50℃下，持续搅拌，乳清蛋白质可以形成浓度大于40%的稳定均匀的溶液，室温下黏度在200~300 cP，pH在6.0~7.0之间。其次，加热到50℃以上，乳清蛋白质分子开始变性展开，同时通过分子间或分子内的巯基-双硫键交换或者巯基-巯基氧化反应交联成大分子多聚体（Monahan et al., 1995; Dunkerly and Zadow, 1984; Haynes and Norde, 1995; Norde and Favier, 1992），当乳清蛋白质胶水固化后，这种热变性形成的大分子网络结构可以产生足够的黏合力。

热处理是非常经济和简单的使乳清蛋白质球蛋白结构展开并交联成大分子聚合物的方法，只要将乳清蛋白质溶液边搅拌边加热至55℃即可。热变性不仅使包埋在球蛋白结构里的极性基团暴露出来，增加了分子间的非共价键作用力（如氢键），也可以促成蛋白的自我交联，使得胶水的内聚力更强。图9.1展示了40%浓度的乳清蛋白质溶液在60℃温度下，黏度随着加热时间的延长而增加。乳清蛋白质溶液的黏度也取决于蛋白的浓度，浓度越高，蛋白交联越多，因此黏度也越高。例如，当蛋白浓度大于15%时，在65℃条件下加热可使蛋白凝胶；但当浓度为10%时，需要加热到85℃并且加热时间为30 min以上才能使溶液凝胶。对40%的乳清蛋白质溶液，最好在60~63℃条件下加热，加热时间不要超过45 min，最好在25~35 min之间。这样处理过的乳清蛋白质黏度和黏合强度最适合用于胶水，参见图9.1和表9.1。乳清蛋白质的黏度随着加热时间的延长而增加；但黏合强度随着加热时间延长达到一个峰值后反而会下降。加热时间过长反而使得黏合强度下降，这可能是因为加热时间过长，乳清蛋白质的黏度过高，使得涂抹难度增加，同时胶水的湿润性降低，导致胶水不易渗入到黏合面的微孔中形成足够的机械互锁。

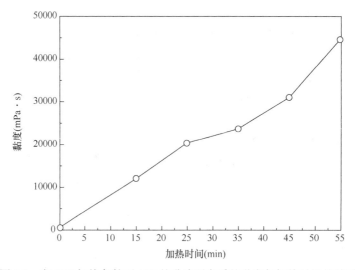

图9.1 在60℃加热条件下40%的乳清蛋白质的黏度与加热时间的关系

表 9.1　60℃加热对 40%的乳清蛋白质溶液作为胶合板胶水的黏合强度的影响

加热时间/min	干黏合强度/MPa	湿黏合强度 A/MPa	湿黏合强度 B/MPa
0（对照）	1.24	1.03	0.69
15	1.65	1.35	0.97
25	1.85	1.56	1.01
35	1.73	1.52	1.08
45	1.57	1.41	0.96
55	1.49	1.09	0.58

注：湿黏合强度 A，黏合样品在 63℃浸泡 3 h 后再测黏合强度。湿黏合强度 B，黏合样品在沸水中煮 4 h，然后在 63℃下干燥 20 h，在沸水中煮 4 h 后再测黏合强度。湿黏合强度 A 和湿黏合强度 B 测定参照 JIS K6806-2003 方法的。

乳清蛋白质胶水的黏合强度可以通过添加含多个活性基团的改良剂来增加交联密度而进一步加强。这些改良剂含有可以与乳清蛋白质交联的活性基团，如乙二醛（GO）、戊二醛（GA）、p-MDI、脲醛树脂、苯酚-甲醛低聚物（PFO）。乙二醛和戊二醛各含有两个醛基，可以与乳清蛋白质的氨基进行交联反应，如图 9.2 方程（1）所示。平均每个 p-MDI 含有 2.7 个异氰基，异氰基也可以与氨基反应[图 9.2 方程（2）]。脲醛树脂和 PFO 含有的大量羟甲基也可以交联乳清蛋白质[图 9.2 方程（3）]。然而表 9.2 的结果显示，这些加了改良剂的乳清蛋白质胶水的湿黏合强度反而没有纯乳清蛋白质胶水高，这说明加了改良剂的胶水的抗湿抗水性反而不如乳清蛋白质本身。

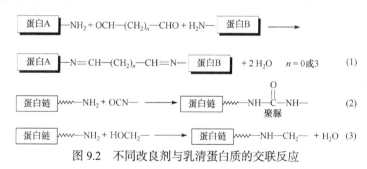

图 9.2　不同改良剂与乳清蛋白质的交联反应

表 9.2　不同改良剂对乳清蛋白质胶合板胶水性能的影响

胶水成分	干黏合强度/MPa	湿黏合强度 B/MPa	甲醛释放浓度/(mg/L)
乳清蛋白质（WP）	1.51	0.98	0.042
WP + 0.15wt% GA	1.98	0.83	0.078
WP + 1wt% GO	1.40	0	0.028
WP + 1wt% p-MDI	1.78	0.93	0.032
WP + 25wt% UF 树脂	1.72	0.67	0.667
WP + 30wt% PFO	1.98	1.73	0.067

注：湿黏合强度 B，黏合样品在沸水中煮 4 h，然后在 63℃下干燥 20 h，在沸水中煮 4 h 后再测黏合强度。参照 JIS K6806-2003 方法。资料来源：Wang et al. (2011)。

乙二醛、戊二醛、p-MDI 对乳清蛋白质的反应活性非常强。添加这些成分将使乳清蛋白质的黏度迅速升高，考虑到最终胶水的湿润性和涂抹性，必须适当减少改良剂的添加量。对于戊二醛和 p-MDI，最佳添加量分别为 0.2%和 1%。如果添加量过高，将导致凝胶过快，以致没有足够的时间将胶水涂抹到木板上。乙二醛在常温下对蛋白的反应不明显，这是因为乙二醛的两个醛基相连，互相形成位阻。乙二醛的最高添加量为 2%，最佳添加量为 1%。表 9.2 结果显示，添加戊二醛和 p-MDI 导致乳清蛋白质的干黏合强度分别增加 31.1%和 17.9%，但是抗湿性和湿黏合强度反而都有轻微地减小。添加乙二醛的乳清蛋白质的干黏合强度比乳清蛋白质略低，但其抗湿性极差，完全无法承受沸水处理，导致湿黏合强度为 0。这可能也是因为乙二醛的两个醛基相连导致互相产生位阻效应，从而降低其与乳清蛋白质的交联反应。脲醛树脂通常是中性和弱碱性（pH7.6 左右），与乳清蛋白质混合并不会导致黏度明显增加。表 9.2 显示，添加 25%的脲醛树脂，可以使乳清蛋白质的干黏合强度增加 13.9%，但是同样，湿黏合强度反而减小 31.6%；另外甲醛释放量高达 0.67 mg/L，远远高于 JIS A5908-2003 所规定的绿色胶合板不能高于 0.3 mg/L 的规定。综上所述，乙二醛、戊二醛、p-MDI、脲醛树脂均不适合作为乳清蛋白质胶水的交联剂用来提高乳清蛋白质胶水的防水抗湿性。

苯酚-甲醛树脂是一种热硬化胶水，但溶解性较差，黏度过高，并且碱性强。如果将苯酚-甲醛树脂在较低温度下（60～75℃）加入少量 NaOH 作为催化剂，可将苯酚-甲醛树脂降解成苯酚-甲醛低聚物（PFO）。PFO 主要含有多羟甲基苯酚和其他低分子量的多聚体，因此 PFO 具有较好的溶解性和较低的黏度，非常适合作为乳清蛋白质的交联剂。在与乳清蛋白质混合前，需要将 PFO 的 pH 调至中性并且添加适量的氨水和亚硫酸钠去除多余的自由甲醛（PFO 中自由甲醛含量 5.1%）。预处理后的 PFO 与乳清蛋白质具有非常好的混合性。在乳清蛋白质溶液中添加 30%的 PFO 可以明显改善乳清蛋白质胶水的黏合强度和防水抗湿性，干湿黏合强度分别为 1.98 MPa 和 1.73 MPa，分别增加 31.1%和 76.5%，也远远高于 JIS K6806-2003 对建筑用胶合板胶水强度的要求（1.18 MPa 和 0.98 MPa）。同时，这种 PFO 乳清蛋白质胶水的甲醛释放量为 0.067 mg/L，添加 PFO 并没有明显增加甲醛释放量（表 9.2），也远远低于 JIS A5908-2003 对绿色胶合板甲醛释放量的要求（＜0.3 mg/L）。用 ASTM D6007 和 D5197 方法测 PFO 乳清蛋白质胶水的甲醛释放量为 0.0043 ppm，远远低于美国加利福尼亚州空气资源协会（CARB）第二阶段排放标准（0.05 ppm）。苯酚是一种比甲醛毒性更强的物质，可以通过皮肤吸收、呼吸道等进入人体造成肌肉无力、抽搐，甚至昏迷（Bruce et al.，1987）。图 9.3 中高效液相色谱（HPLC）分析显示苯酚的峰出现在 19.40 min，无论是硬化前还是硬化后的 PFO 乳清蛋白质胶水均没有检测到苯酚。这也证实了经过预处理的 PFO 已经不含自由苯酚。综合考虑 PFO 乳清蛋白质胶水的黏合强度、防水抗湿性及其环保安全性等各方面因素，用 PFO 交联乳清蛋白质将具有很高的潜在商业应用价值。

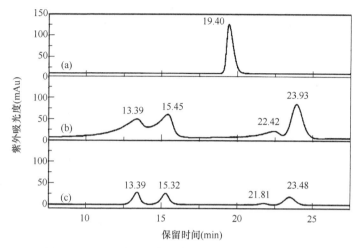

图 9.3　HPLC 检测乳清蛋白质胶水中的自由苯酚含量
（a）苯酚溶液；（b）未硬化胶水；（c）硬化后的胶水。资料来源：Wang et al.（2012）

除了在胶合板胶水中的应用，乳清蛋白质还可以用作冷压胶合木（glulam）胶水。用于胶合木的胶水包括 RF 树脂、多异氰酸酯聚合物树脂、水性异氰酸酯聚合物树脂（API）。API 胶水含有两个组分，包括水溶性胶水和异氰酸酯交联剂。水溶性胶水可以是聚乙烯醇（PVA）溶液、丁苯橡胶（SBR）乳化液、乙烯/乙酸乙烯酯（EVA）共聚物乳化液，或者它们之间的混合物（Hori et al.，2008）。异氰酸酯交联剂是由 p-MDI 粗料形成的。热变性的乳清蛋白质溶液也同样适用于基于乳清蛋白质的 API 胶水。

结构用胶合木胶水的强度要求要远远高于胶合板胶水的强度，根据 JIS K6806-2003 标准，结构用胶水的最低强度要求是 9.81 MPa。从表 9.3 看出，乳清蛋白质本身的黏合强度远远低于这个要求。同时表 9.3 中列出的配方根本无法承受 28 h 沸水-干燥的测试，胶水的耐力也远远达不到要求，因此势必需要一种更强的交联剂。

表 9.3　不同乳清蛋白质胶水配方的性能比较

胶水编号和配方	工作寿命/h	黏合强度/MPa	
		干黏合强度	湿黏合强度
A1: 乳清蛋白质（WP）溶液	N/A	2.06	≈ 0
A2: 100 g WP，15 g p-MDI	0.5	5.78	2.64
A3: 70 g WP，30 g PVAc，15 g p-MDI	2.3	6.02	3.70

续表

胶水编号和配方	工作寿命/h	黏合强度/MPa	
		干黏合强度	湿黏合强度
A4: 58.3 g WP, 11.7 g PVA, 30 g PVAc, 15 g p-MDI	2.0	10.56	5.65
A5: 55.4 g WP, 11.1 g PVA, 30 g PVAc, 15 g p-MDI, 3.5 g $CaCO_3$	2.1	13.38	6.81
A6: 市面 API 胶水	2.8	12.98	6.37

注：湿黏合强度 B，黏合样品在沸水中煮 4 h，然后在 63℃下干燥 20 h，在沸水中煮 4 h 后再测黏合强度。参照 JIS K6806-2003 方法。

乳清蛋白质富含的羟基（0.11 mol/100 g 乳清蛋白质）和氨基残基（0.13 mol/100 g 乳清蛋白质）对 p-MDI 交联剂的反应活性很强（McDonough et al., 1974），所以用 p-MDI 交联的乳清蛋白质胶水的强度和耐久性会大大增强。乳清蛋白质和 p-MDI 的交联反应可以分为三个反应步骤（图 9.4），并且交联在常温 20~25℃条件下就能进行。这种高强度的交联反应可以防止胶水硬化后在高温高湿的环境下被破坏，同时这种高强度的交联反应也产生胶水内部分子强大的内聚力，更加增强了胶水的黏合强度和耐受性。异氰酸酯基团同时也可以和木纤维素中的羟基交联形成氨基甲酸酯键（化学键），这更加增强了胶水和黏合面之间的黏合强度，使得胶面更强固。表 9.3 列出了各种不同乳清蛋白质胶水的配方及其黏合强度和耐受力。配方 A2 采用了 100 g 热变性的乳清蛋白质溶液和 15 g p-MDI，其黏合强度和耐受力与 A1（无交联剂）相比显著增强，甚至在承受 28 h 沸水-干燥测试后仍有 2.64 MPa 的湿黏合强度。

但是 A2 配方的工作寿命很短，只有半小时左右。乳清蛋白质和交联剂混合后，会迅速变得非常黏稠，并且生成颗粒状的聚集物。这是因为蛋白质中的氨基和交联剂的异氰酸酯基团迅速反应生成不溶性的聚脲链，迅速使分子量不均匀增加，从而形成颗粒状的聚集物[图 9.4 方程（1）]。在 p-MDI 交联剂中先添加 30% 的 PAVc（液体）混合，可以有效地减缓蛋白交联速度，从而使得胶水的工作寿命达 2.3 h，湿黏合强度提高到 3.70 MPa（表 9.3 配方 A3）。如果把 PVAc 事先与乳清蛋白质溶液混合，再加入交联剂，可以进一步提高工作寿命至 3.6 h，但是湿黏合强度降低到 1.32 MPa。这个结果显示，乳清蛋白质、PVAc、p-MDI 不同的混合顺序对最终 API 胶水的性能也有影响（Zhao et al., 2011）。

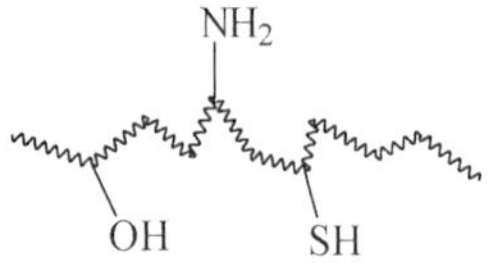

乳清蛋白质链 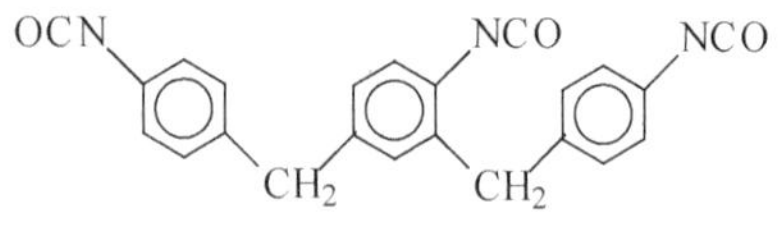　　　　　　　　典型的p-MDI结构

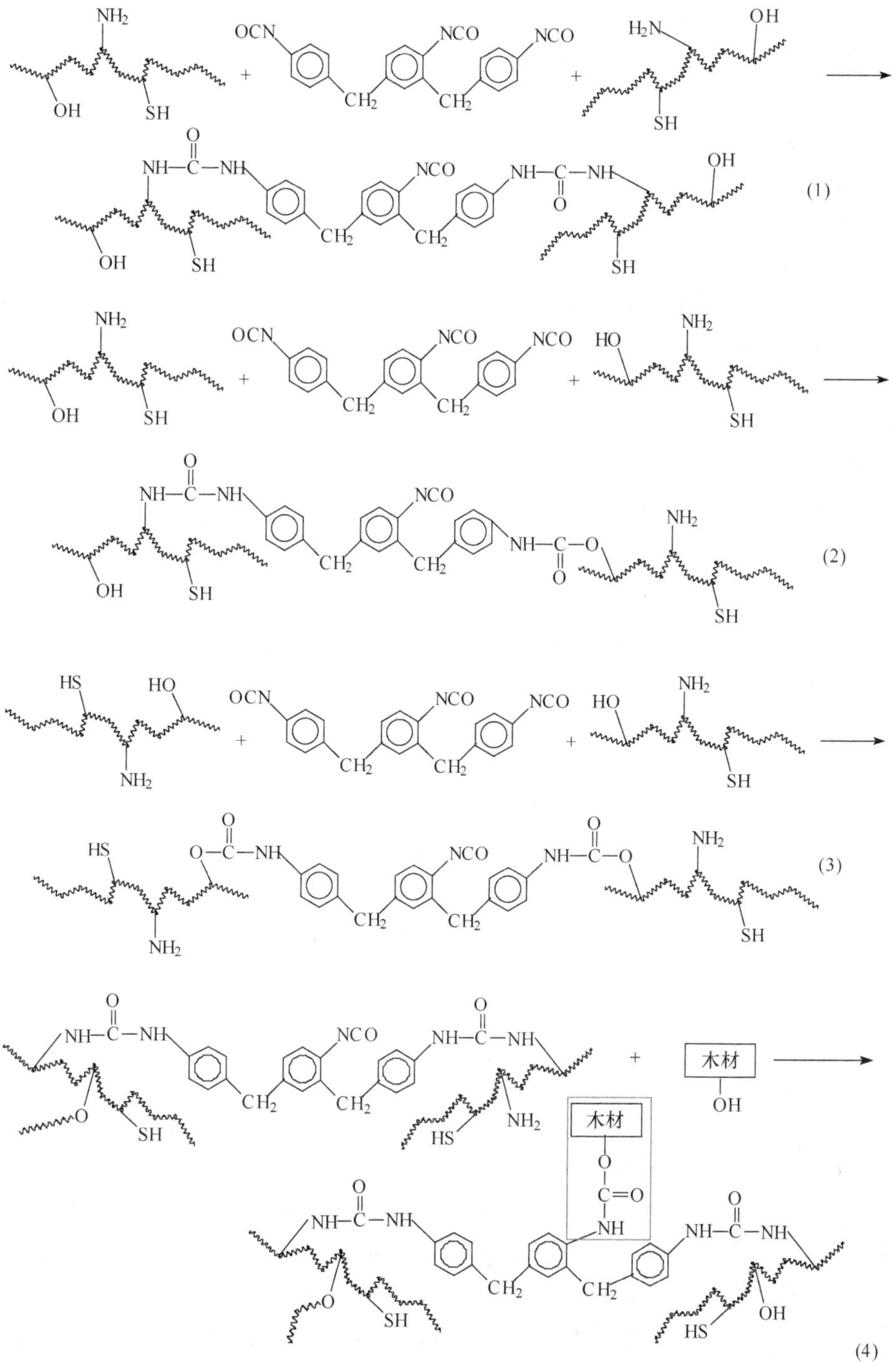

图 9.4　乳清蛋白质 API 胶水的反应原理

PVA 分子是含有—CH_2—CH(OH)—重复单位的高分子聚合物,含有大量的羟基基团可以与 p-MDI 反应。PVA 可以与乳清蛋白质交联（Lacroix et al., 2002; Srinivasa et al., 2003),因此在配方 A3 的基础上加入 11.7%的 PVA 形成配方 A4（表 9.3）。A4 的干黏合强度可达 10.56 MPa,湿黏合强度可达 5.65 MPa,分别比配方 A3 增加了 75.4%和 52.7%。这也证实了 PVA 可以有效增强乳清蛋白质和 p-MDI 胶水的强度。但即使加了 PVA 后,胶水的湿黏合强度仍然低于法规要求的 5.88 MPa 以上。

一些纳米级的填充物如 SiO_2、Al_2O_3、$CaCO_3$ 可以进一步帮助提升胶水的耐久力,这些填充物可以增加胶水与黏合面的接触面积从而增加胶水和黏合物之间的机械互锁作用（Hussain et al., 1996; Chen et al., 2004; Gilbert et al., 2003）。添加 3.5%的纳米 $CaCO_3$ 之后,胶水的黏合强度和耐久性都显著增加。表 9.3 中配方 A5 的湿黏合强度已达 6.81 MPa,超过了 JIS K6806-2003 标准规定的 5.88 MPa,比没有加纳米 $CaCO_3$ 的配方 A4 强度增加 20.5%。A5 的干黏合强度为 13.38 MPa,远高于 JIS 标准规定的 9.81 MPa。比配方 A4 强度（10.56 MPa）增加了 26.7%（Gao et al., 2011）。配方 A5 的干黏合强度和湿黏合强度也都好于市售 API 胶水配方 A6。这显示了乳清蛋白质 API 胶水在结构性胶合木胶水中也有非常好的应用前景和潜力。

9.4 办公胶水

办公胶水并没有非常明确的定义,一般泛指所有在学校、办公室、家用于粘贴纸张或其他物品的胶水,通常也称为纸胶水。办公胶水通常有液体胶和固体胶两大类。目前绝大多数办公胶水的材料是合成的聚合物材料,如聚乙烯醇（PVA）、聚乙酸乙烯酯（PVAc）和聚乙烯吡咯烷酮（PVP）。安全性是办公胶水的一个重要指标,因为办公胶水会经常与皮肤接触,甚至会被小孩误食。在合成胶水问世之前,淀粉胶水是最主要的天然原料的胶水。乳制品,如酪蛋白,也广泛运用在胶水中。但乳清蛋白质在办公胶水中的应用还比较少。

蛋白质大多是天然的胶水原料,因为蛋白质分子含有大量的羟基,可以与木或纸中纤维素的羟基形成氢键,从而产生黏合强度。同时蛋白质分子会随着水分渗入到纸的内部,在水分蒸发胶水硬化后形成机械互锁力。但乳清蛋白质是高度折叠紧凑的球蛋白,并且分子量不是很大,这种结构本身并不适于作为胶水的原料,因为胶水原料通常需要大分子的线形聚合体;乳清蛋白质的另外一个缺陷是黏度太低,一般液体胶水需要一定的黏度以便在胶水硬化前抓住纸张。因此,必须对乳清蛋白质进行适当的改性使之更适合用于胶水（Guo and Wang, 2016 a, b）。

热处理是打开乳清球状蛋白结构最常用的方法。球蛋白结构破坏后，埋藏在内部的活性基团如巯基就会暴露出来，发生巯基-双硫键交换反应，将分子间的双硫键变成分子内的双硫键，使得小分子的球蛋白形成高分子的聚合物（Guo and Wang，2016 a，b）。图 9.5 显示了 10%的乳清蛋白质溶液在不同温度下加热 10 min 后的黏度。如果乳清蛋白质溶液的黏度小于 100 mPa·s，用作胶水就会太稀，无法有效粘住纸片。乳清蛋白质在 90℃下加热 10 min 后就会形成有黏性的黏稠液体（Wang et al.，2013）。除了加热以外，其他凡是能暴露乳清蛋白质的巯基基团，并且促成巯基-双硫键交换反应的方法，如改变溶液极性和 pH 等，都可能有助于将乳清蛋白质变成适合应用于胶水的原料（van der Leeden et al.，2000）。

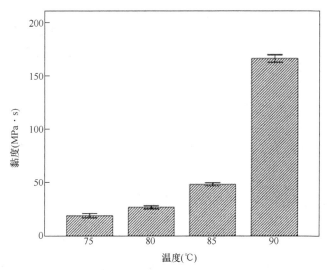

图 9.5　在不同温度下加热 10 min 对乳清蛋白质溶液（10%）黏度的影响
资料来源：Wang et al.（2013）

早在 1953 年，Tschabold 和 Mueller 就用乳清为原料做成了一种纸胶水，并且申请了专利。他们用干酪生产中直接排出的乳清（含 4%～5%乳糖和 0.6%蛋白质），中和 pH 后，加热将乳清浓缩到 5∶1 后形成黏稠的液体，再按 2∶1 添加糊精。这样制成的胶水即使产品冷却后也有黏性。

乳清蛋白质作为胶水原料的主要优势是其安全性（其本身作为一种食品原料）。除了安全和黏合强度大之外，办公胶水还需要具有使用方便、保质期长等特点。为了达到合适的黏度和黏合强度，乳清蛋白质需要经过热变性交联成多聚体（Van der Leeden et al.，2000；Tschabold and Mueller，1953）。但是热变性后的乳清蛋白质在冷却后黏度会一直增加，最终形成硬的凝胶从而丧失胶水的黏性。如何保证变性后的乳清蛋白质有一个稳定的高黏度是研发有商业潜能的乳清蛋白质办公胶水的关键。

热变性后的乳清蛋白质的巯基等活性基团充分暴露,当乳清蛋白质的浓度达到一定水平后,这些活性基团在常温或低温下可以持续反应交联,使胶水黏度缓慢持续增加,最终形成硬化丧失黏性。加入一定的合成胶水原料可以帮助增强乳清蛋白质胶水黏性同时稳定溶液的黏度。这些合成胶水原料可以在乳清蛋白质分子之间形成物理障碍,阻止或者减缓乳清蛋白质分子间的相互作用力。PVA、PVAc、PVP 等都是在办公胶中广泛使用的胶水原料。但是 PVA 和 PVAc 的分子仍具有极性或弱极性,可以与乳清蛋白质的羟基形成氢键,并不能有效稳定乳清蛋白质的黏度。将 PVA 和 PVAc 与乳清蛋白质混合后,在两个月之内胶水也会硬化而丧失黏性(Wang et al., 2013)。PVP 由于支链结构的不同,不能交联蛋白(图 9.6),因此可以用来稳定变性乳清蛋白质的黏度。蔗糖也可以有效降低乳清蛋白质的冷凝胶速度(Kulmyrzaev et al., 2000)。另外,蔗糖作为一种非还原糖,不会与蛋白反应。因此,蔗糖也是一种合适的变性乳清蛋白质胶水的黏度稳定剂。

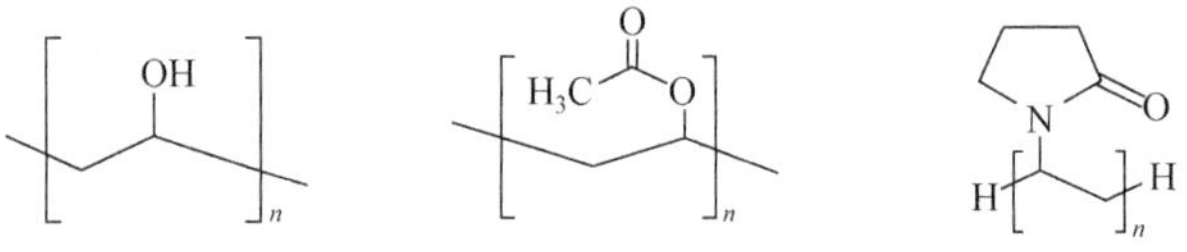

图 9.6 聚乙烯醇(PVA)(a)、聚乙烯乙烯酯(PVAc)(b)和聚乙烯吡咯烷酮(PVP)(c)的分子结构

研究显示,蔗糖和 PVP 可以用来与变性乳清蛋白质混合制成稳定的液体办公胶水(Wang et al., 2013;Wang and Guo, 2014)。工艺流程如下:先将 10%的乳清蛋白质溶液在 75℃以上的温度加热并变性,然后加入 PVP 溶液或者蔗糖,同时加入抑菌剂。这样制成的胶水可以常温或 37℃保存一年以上不会凝胶或丧失黏性。表 9.4 列出了 WPI/PVP 和 WPI/蔗糖胶水的黏合强度和黏度。其黏合强度均达到市场上销售的商业办公胶水水平(Wang et al., 2013;Wang and Guo, 2014)。

表 9.4 乳清蛋白质办公胶水的黏度和黏合强度

产品	黏度(mPa·s)	黏合强度(MPa)
胶水 A: WPI 7.5% + PVP 7.5%	918 ± 212	1.38 ± 0.02
胶水 B: WPI 12% + 蔗糖 10%	3090 ± 300	1.42 ± 0.03
市面产品对照	3692 ± 363	1.46 ± 0.09

资料来源:Wang et al. (2013);Wang and Guo (2014)。

胶棒是另外一种形式的办公胶。通常是用硬脂酸钠固化 PVA 或者 PVP 液体胶水。蔗糖和甘油通常用来当作填充剂和保湿剂。跟液体胶水一样,如何长时间保存变性乳清蛋白质胶水的黏性、防止其失水硬化是关键。蔗糖和PVP同样可以对胶棒中的乳清蛋白质起到保护作用。将WPI、PVP、蔗糖、保湿剂、消泡剂和抑菌剂溶解在水中并混合，加热至90℃并保持20 min，然后将融化的热胶水倒入胶棒模具或胶棒盒中冷却定型（Wang et al., 2012）。乳清蛋白质胶棒的配方和黏合强度均列在表9.5中。这种胶水在常温和37℃高温下可以放置一年以上而不丧失黏性。

表9.5 乳清蛋白质胶棒和PVP胶棒的配方及产品性能对照

		乳清蛋白质胶棒	PVP 胶棒	市面产品对照
配方	WPI 粉	8%	0%	—
	PVP 粉	8%	16%	—
	蔗糖	20%	20%	—
	甘油	10%	10%	—
	硬脂酸钠	7%	7%	—
性能	黏合强度（kPa）	1355 ± 89	1221 ± 115	1300 ± 192
	胶棒硬度（N）	20.11 ± 0.64	55.45 ± 5.47	21.14 ± 1.82

资料来源：Wang et al.（2012）。

9.5 手术胶

手术胶是在1950年出现的一种新型的手术缝合技术，特别是在1980年当第一款手术胶在欧洲和加拿大获批并在临床中应用后，这项技术得到迅速发展和应用。Quinn（2005）系统地总结了手术胶的应用和发展现状。在美国有一款FDA批准并成功应用的手术胶 BioGlue®（图9.7）。BioGlue®是一种典型的复合胶水：45%的牛乳清白蛋白（BSA）和少量的戊二醛（GTA）。BSA和GTA在管尖混合，BSA被GTA交联（通过游离氨基和醛基反应），注射后肌肉细胞中的ε-氨基通过GTA与BSA交联，这个反应非常迅速，通常只需几秒钟，从而对伤口进行黏合（图9.8）。BioGlue®使用的BSA，也是乳清蛋白质中的第三大蛋白组分。BSA与其他乳清蛋白质组分类似，都是球蛋白并且含有丰富的ε-氨基。大量的ε-氨基可以让乳清蛋白质和BSA一样提供足够多的交联点，产生足够大的黏合强度。这些性质都表明了乳清蛋白质在手术胶中使用的可能性。

图9.7 BioGlue®产品图

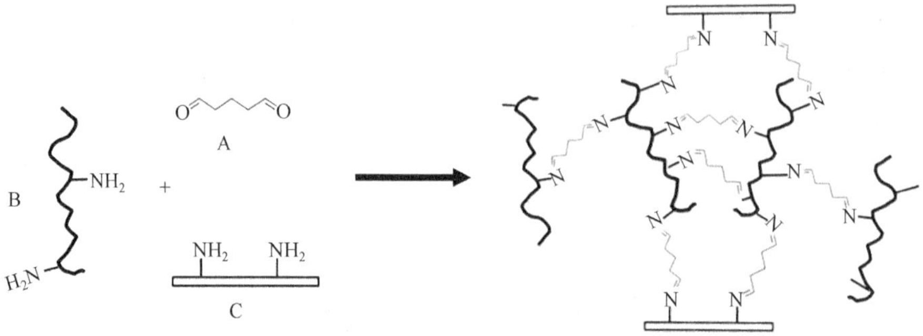

图 9.8　戊二醛与蛋白质交联反应及手术胶黏合原理
A. 戊二醛；B. 蛋白胶水；C. 组织蛋白

9.6　总　　结

由于乳清蛋白质含有丰富的功能性基团，如氨基、羟基、羧基，这些基团可以通过改性和交联来增加分子量并且形成三维网状结构。这使得乳清蛋白质可以被用来制作许多环保型的胶水类产品。与其他蛋白相比，乳清蛋白质的主要优点是热变性和良好的溶解性。热变性的稳定网状交联结构使得乳清蛋白质胶水具有非常好的黏合力和耐久力。除了无毒、无挥发的优点外，乳清蛋白质作为一种可降解的蛋白质，可以有效地降低甚至完全避免有毒合成材料在胶水中的使用，是一种非常环保的胶水材料。

尽管目前对于乳清蛋白质在胶水中的应用的研究还不多，但是目前已经证明了技术上的可行性，应用的障碍主要是成本而非技术。乳清蛋白质作为一种优质蛋白在运动营养和婴幼儿配方粉中的需求仍然很大。但是乳清蛋白质作为胶水的应用还是有机会的，如在一些高附加值的黏合剂（如手术胶）中的使用具有很好的前景。

参 考 文 献

Audic, J.L., Chaufer, B., and Daufin, G. (2003). Non-food applications of milk components and dairy co-products: a review. *Le Lait* **83** (6): 417-438.

Bruce, R., Santodonato, J., and Neal, M. (1987). Summary review of the health effects associated with phenol. *Toxicology and Industrial Health* **3** (4): 535-568.

Chen, H., Sun, Z., and Xue, L. (2004). Properties of nano SiO2 modified PVF adhesive. *Journal of Wuhan University of Technology-Mater. Sci. Ed.* **19**: 73-75.

Dunkerly, J.A. and Zadow, J.G. (1984). The effect of calcium and cysteine hydrochloride on the firmness of heat coagula formed from cheddar whey protein concentrates. *Australian Journal of Dairy Technology* **39**: 44-47.

Gao, Z., Wang, W., Zhao, Z. et al. (2011). Novel whey protein-based aqueous polymer-isocyanate adhesive for glulam. *Journal of Applied Polymer Science* **120** (1): 220-225.

Gao, Z., Yuan, J., and Wang, X. (2007). Phenolated larch-bark formaldehyde adhesive with multiple additions of sodium hydroxide. *Pigment & Resin Technology* **36** (5): 279-285.

Gilbert, E.N., Hayes, B.S., and Seferis, J.C. (2003). Nano-alumina modified epoxy based film adhesives. *Polymer Engineering and Science*. **43**: 1096-1104.

Guo, M. and Wang, G. (2016a). Whey protein polymerisation and its applications in environmentally safe adhesives. *International Journal of Dairy Technology* **69** (4): 481-488.

Guo, M. and Wang, G. (2016b). Milk protein polymer and its application in environmentally safe adhesives. *Polymers* **8** (9): 324.

Haynes, C.A. and Norde, W. (1995). Structure and stabilities of adsorbed protein. *Journal of Colloid Interface Science* **169**: 313-328.

Hemingway, R.W., Conner, A.H., and Branham, S.J. (eds.) (1989). *Adhesives from Renewable Resources*, ACS Symposium Series 385. Washington, DC: American Chemical Society.

Henderson, J.T. (1979). Volatile emissions from the curing of phenolic resins. *Tappi Journal* **62**: 93-96.

Hori, N., Asai, K., and Takemura, A. (2008). Effect of the ethylene/vinyl acetate ratio of ethylene-vinyl acetate emulsion on the curing behavior of an emulsion polymer isocyanate adhesive for wood. *Journal of Wood Science* **54**: 294-299.

Hussain, M., Nakahira, A., and Niihara, K. (1996). Mechanical property improvement of carbon fiber reinforced epoxy composites by Al_2O_3 filler dispersion. *Materials Letters* **26**: 185-191.

Kulmyrzaev, A., Cancelliere, C., and McClements, D.J. (2000). Influence of sucrose on cold gelation of heat-denatured whey protein isolate. *Journal of the Science of Food and Agriculture* **80** (9): 1314-1318.

Lacroix, M., Le, T.C., Ouattara, B. et al. (2002). Use of gamma-irradiation to produce films from whey, casein and soy proteins: structure and functional characteristics. *Radiation Physics and Chemistry* **63**: 827-832.

van der Leeden, M.C., Rutten, A.A., and Frens, G. (2000). How to develop globular proteins into adhesives. *Journal of Biotechnology* **79** (3): 211-221.

McDonough, F., Hargrove, R., Mattingly, W. et al. (1974). Composition and properties of whey protein concentrates from ultrafiltration. *Journal of Dairy Science* **57**: 1438-1443.

Meyer, B. and Hermans, K. (1986). Formaldehyde release from wood products: an overview. In: *Formaldehyde Release from Wood Products*, 1-16. Washington, DC: American Chemical Society.

Monahan, F.J., German, J.B., and Kinsella, J.E. (1995). Effect of pH and temperature on protein unfolding and thiol/disulfide interchange reactions during heat-induced gelation of whey proteins. *Journal of Agriculture and Food Chemistry* **43**: 46-52.

Norde, W. and Favier, J.P. (1992). Structure of adsorbed and desorbed proteins. *Colloids and Surfaces* **64** (1): 87-93.

Perera, F. and Petito, C. (1982). Formaldehyde: a question of cancer policy? *Science* **216**: 1285-1291.

Pocius, A.V. (2012). *Adhesion and Adhesives Technology*, 2e. Munich, Germany: Hanser.

Quinn, J.V. (2005). *Tissue Adhesives in Clinical Medicine*. Hamilton, Canada: BC Decker.

Srinivasa, P.C., Ramesh, M.N., Kumar, K.R. et al. (2003). Properties and sorption studies of chitosanpolyvinyl alcohol blend films. *Carbohydrate Polymers* **53**: 431-438.

Stoeckel, F., Konnerth, J., and Gindl-Altmutter, W. (2013). Mechanical properties of adhesives for bonding wood - a review. *International Journal of Adhesion and Adhesives* **45**: 32-41.

Swenberg, J.A., Kerns, W.D., Mitchell, R.I. et al. (1980). Induction of squamous cell carcinomas of the rat nasal cavity by inhalation exposure to formaldehyde vapor. *Cancer Research* **40**: 3398-3402.

Tschabold, G.L. and Mueller, D.L. (1953). Adhesive from whey and a method of making it. U.S. Patent 2,624,679, issued January 6 1953.

USDA Forest Products Laboratory (1987). *Wood Handbook: Wood as an Engineering Material*. Ag. Handbook 72 (Rev). Madison, WI: Forest Products Laboratory.

Wang, G., Cheng, J., Zhang, L. et al. (2012). Physicochemical and functional properties, microstructure, and storage stability of whey protein/polyvinylpyrrolidone based glue sticks. *BioResources* **7** (4): 5422-5434.

Wang, G. and Guo, M. (2014). Property and storage stability of whey protein-sucrose based safe paper glue. *Journal of Applied Polymer Science* https://doi.org/10.1002/app.39710.

Wang, G., Zhang, T., Ahmad, S. et al. (2013). Physicochemical and adhesive properties, microstructure and storage stability of whey protein-based paper glue. *International Journal of Adhesion and Adhesives* **41**: 198-205.

Wang, W., Zhao, Z., Gao, G. et al. (2011). Whey protein-based water-resistant and environmentally safe adhesives for plywood. *BioResouces* **6** (3): 3339-3351.

Wang, W., Zhao, Z., Gao, Z. et al. (2012). Water-resistant whey protein based wood adhesive modified by post-treated phenol-formaldehyde oligomers (PFO). *BioResources* **7** (2): 1972-1983.

Zhao, Z., Gao, Z., Wang, W. et al. (2011). Formulation designs and characterisations of whey-protein based API adhesives. *Pigment & Resin Technology* **40** (6): 410-417.

第 10 章　乳清蛋白质展望

Mingruo Guo[1,2] and Guorong Wang[1]

1. Department of Nutrition and Food Science, University of Vermont, Burlington, USA
2. College of Food Science, Northeast Agriculture University, Harbin, People's Republic of China

10.1　乳清蛋白质市场需求

为了能更全面地模拟人乳，企业在生产婴儿配方奶粉时常用乳清蛋白质和乳糖来调整牛乳中乳清蛋白质与酪蛋白、乳蛋白与乳糖的比例。随着全球范围内中产阶层职场母亲人数的不断增加，婴幼儿配方奶粉市场以每年 6%左右的速度增长，其中，中国中产阶层人数的激增起到了主要的推动作用。2016 年，中国新生婴儿人数增加了 131 万（Levenson，2017）。此外，下一代婴幼儿配方奶粉开发将更加侧重于模拟母乳中的特定蛋白质，而不仅仅是调整乳清蛋白质与酪蛋白的比例，对富含如α-LA、Lf 的乳清蛋白质的需求会持续增长。

乳清蛋白质对肌肉力量、抵抗力和修复有积极作用，因而在运动营养方面有较好的应用前景（Ha and Zemel，2003；Hayes and Cribb，2008）。在运动营养领域源于美国的"乳清热"，现在已经火遍全球。根据市场研究公司 Zion Market Research 发布的一份报告，2016 年全球运动营养市场规模约为 283.7 亿美元，预计到 2022 年将达到 452.7 亿美元，年增长率为 8.1%。虽然运动营养是一个新兴市场，但预计 2022 年其市值或将接近 2016 年的婴幼儿配方奶粉市值（约 502 亿美元）。由于母乳喂养能促进母亲和婴儿健康，婴幼儿配方奶粉的增长速度可能不如运动营养食品快，其最大乳清蛋白质市场的地位也可能被运动营养取代。

10.2　希腊酸奶的盛行和酸乳清

目前，几乎所有商品乳清浓缩蛋白和乳清分离蛋白都是从甜乳清中提取回收的。酸乳清仍主要用于饲料（Schingoethe，1976）、肥料（Jones et al.，1993）、生物转化（Mawson，1994）和酸乳清粉生产等方面（Nassauer et al.，1996）。酸乳清的体量比甜乳清小得多，所以未引起学术界和食品工业界的过多关注。近年来，希腊酸奶的产销量从美国开始并在世界范围内迅速增长（Bieldt，2013；Gurel，2016）。与传统酸奶相比，希腊酸奶弃去液体酸乳清（占总体积2/3）浓缩成固态，具有高蛋白、高钙和低乳糖等特点，其生产过程如图10.1所示。希腊酸奶的蛋白质和钙含量是传统酸奶的三倍多，乳糖含量比传统酸奶少1/3。因此，希腊酸奶能够吸引那些有高蛋白饮食或减肥代餐需求的消费者（Dharmasena et al.，2014）。2010~2015年，希腊酸奶在美国的销售额从3.91亿美元增长到37亿美元，而非希腊酸奶的销售额则从44亿美元下降到40亿美元。

牛奶→巴氏杀菌法→发酵→酸乳清分离→希腊酸奶
↓
酸乳清

图10.1 希腊酸奶生产工艺

另一种希腊酸奶的生产工艺是添加其他不同蛋白质来降低酸乳清含量以提高产品蛋白质含量（Peng et al.，2009；Bong and Moraru，2014），但这种工艺所使用的复水蛋白质会导致产品的口感、质地和味道发生劣变，如沙质感、粗糙感等。随着希腊酸奶市场规模的不断扩大，酸乳清的利用正受到越来越多的关注，近几十年来，蛋白质的回收、脱盐技术不断进步和甜乳清成熟的应用技术都推动了酸乳清利用技术的发展。

因为酸含量高，乳清粉非常黏稠，所以酸乳清的干燥是难以解决的问题（Keller，2013），碱中和或加入脱脂粉可以解决（Keller，2013），但是加碱会增加灰分含量，并对酸乳清粉的营养和功能性产生不利影响，目前纳滤和电透析技术替代了中和方法去除乳酸和乳酸盐（Chandrapala et al.，2016；Chen et al.，2016）。虽然膜技术已经成熟应用于甜乳清的生产加工，在酸乳清浓缩蛋白和酸乳清分离蛋白的生产中也开始应用（Zydney，1998），但由于酸乳清与甜乳清的组成有差异，特别是钙和磷酸盐的差异很大，膜技术在酸乳清加工中的应用受到限制（Chandrapala et al.，2015）。酸奶发酵过程中的酸溶作用使酪蛋白微粒中的钙溶出，高钙浓度降低了蛋白质的溶解性并引起乳清蛋白质凝聚，显著提高了膜污染程度（Chandrapala et al.，2015）。将酸乳清pH调至7.0，并在膜过滤前离心澄清，能显著提高膜通量（Kuo and Cheryan，1983）。

酸乳清蛋白质组分的分级分离利用是酸乳清高附加值开发的另一途径。市场对乳清浓缩蛋白和乳清分离蛋白的需求量很大，但仅有少量的甜乳清能用于生产这两种分级蛋白。乳清蛋白质中的α-LA、Lf具有优良的营养性和功能性，也是人乳中含量最高和第二的蛋白质，β-LG有优良的凝胶性，能用于酸奶等产品加工。研究表明，利用水双相体系（Kalaivani and Regupathi，2015）和超滤/沉淀方法

（Muller et al.，2003）可从酸乳清中提取α-LA 和β-LG。含量较少的乳清蛋白质组分如 Lf 可通过羧甲基阳离子交换色谱（Yoshida，1991）和磁吸分离法（Chen et al.，2007）从酸乳清中分离出来。Lf 在牛乳中含量较少，但在人乳中含量较高，富含 Lf 的婴幼儿配方奶粉可以显著促进婴幼儿身体健康（Chierici et al.，1992；Roberts et al.，1992；Davidsson et al.，1994；Tomita et al.，2002）。尽管 Lf 的需求量和价格都很高，但许多甜乳清生产商却无意生产，他们顾虑的是虽然从 Lf 产品中能够获得利润，但不含 Lf 的乳清蛋白质产品会失去竞争力。因此，利用低值的酸乳清获取 Lf 是解决这个顾虑的办法之一。

10.3　微滤乳和清蛋白

超滤和微滤已广泛应用于乳清蛋白质的生产。超滤的孔径大小介于乳清蛋白质分子和乳糖分子之间，能过滤乳糖截留回收乳清蛋白质，即可从液体乳清中提取乳清蛋白质。超滤能同时截留酪蛋白和乳清蛋白质用于生产乳浓缩蛋白和分离蛋白。微滤的孔径相对更大，能透过乳清蛋白质截留如脂肪球和变性蛋白等较大的粒子，在乳清分离蛋白生产中用于脱脂。牛乳中的酪蛋白以胶束形式存在，平均粒径 90～130 nm，比乳清蛋白质粒径大 10～100 倍（Holt，1975）。孔径适当的微滤膜可将酪蛋白微粒和清蛋白分离（Hurt et al.，2010），孔径 100 nm 的微滤膜能将 95%以上的清蛋白从脱脂乳中分离出来（Hurt et al.，2010），这种方法可浓缩原料乳中的酪蛋白以便加工干酪和希腊酸奶（Nelson and Barbano，2005；Bong，2013）。此工艺产生两种乳清产品：微滤牛乳分离出的清乳清和凝乳产生的甜或酸乳清（产量会减少）。与甜乳清蛋白质相比，清蛋白有以下优势。

（1）无添加。干酪生产中添加的凝乳酶会残留在乳清产品中，还有色素、风味剂和其他添加剂都可能影响和污染乳清产品。

（2）风味正。干酪生产中的酶促反应会产生大量风味化合物等成分，导致不同批次的乳清蛋白质风味不一致，产品感官质量不稳定。

（3）纯天然。清蛋白加工工序最少（比其他乳清加工少巴氏杀菌工序），最接近天然牛乳，具有最佳的营养和功能性。

（4）无糖巨肽。糖巨肽是凝乳过程中产生的酪蛋白片段，含量可达甜乳清蛋白质总量的 20%。当乳清蛋白质用于婴幼儿配方奶粉生产时，糖巨肽会改变乳清蛋白质氨基酸组成，使其与母乳成分产生差异。清蛋白是天然的无糖巨肽乳清蛋白质，益于在婴幼儿配方奶粉中应用。

目前，清蛋白是最好的乳清蛋白质。但牛乳中酪蛋白和乳清蛋白质的比例是 4∶1，意味着生产 1 lb（1 lb=0.453592 kg）清蛋白就会产生 4 lb 酪蛋白。以脱脂乳为原料，利用微滤技术生产清乳蛋白已无技术问题。图 10.2 表示了微滤法从脱脂乳中提取清蛋白和酪蛋白微粒的过程。

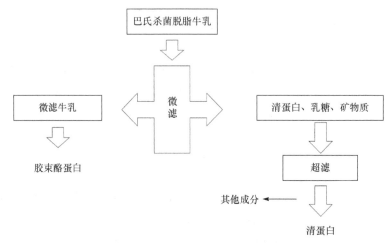

图 10.2　脱脂乳微滤制备清乳清和酪蛋白微粒的流程图

商品酪蛋白主要包括酸性酪蛋白、酪蛋白盐和酶凝酪蛋白，这些产品在加工过程中经过了剧烈的化学处理和酶处理（Savello et al., 1989）。酪蛋白的营养性和功能性如高钙和热稳定性好，部分原因源于其具有的胶束结构，而三种主要的酪蛋白产品（酪蛋白盐、酸性酪蛋白和酶凝酪蛋白）的胶束结构受到一定程度的破坏，导致溶解性降低（酸性酪蛋白和酶凝酪蛋白），或钙含量降低（酪蛋白钠盐和酪蛋白钾盐）。酪蛋白的胶束结构是疏水性微量元素理想的纳米封装的载体（Semo et al., 2007）。胶束酪蛋白已成功应用于希腊酸奶和高蛋白饮料加工（Amelia and Barbano, 2013; Bong and Moraru, 2014）。生产普通酪蛋白会产生大量低质酸乳清（尤其是生产酸性酪蛋白和酪蛋白盐时），如果胶束酪蛋白部分替代目前普通酪蛋白，则可用生产胶束酪蛋白得到的高品质清蛋白取代低质酸乳清，这将会带来巨大的经济和社会效益。

胶束酪蛋白具有优良的热稳定性、钙含量高和有利于肌肉合成等特点，在运动营养中有应用价值（Burd et al., 2012）。酪蛋白和乳清蛋白质都是营养价值高的优质蛋白质，两者在功能性和理化性方面的差异如下。

（1）中性 pH 下胶束酪蛋白对热稳定，乳清蛋白质对热敏感。胶束酪蛋白适宜生产饮料，乳清蛋白质适宜加工成蛋白粉。

（2）胶束酪蛋白对 pH 敏感，乳清蛋白质对 pH 稳定。胶束酪蛋白主要用于甜饮料，乳清蛋白质可用于不同 pH 的饮料。

（3）胶束酪蛋白在胃中消化排空速度较慢，有较好的饱腹感。乳清蛋白质吸收速度较快（Lacroix et al., 2006; Abou-Samra et al., 2011），是运动员体内蛋白质快速消耗后理想的补充剂。对于不活动、选择减肥代餐的人来说，胶束酪蛋白较为理想（能持续提供人体所需氨基酸）（Lacroix et al.，2006）。

10.4 总　　结

运动营养中的"乳清热"和婴幼儿配方奶粉的巨大市场推动了乳清蛋白质市场的发展，以甜乳清为原料的商品乳清蛋白质产品已不能满足需求，希腊酸奶市场规模的不断扩大提高了原料乳清供应量，酸乳清经过适当加工处理或许会弥补乳清蛋白质市场的供需缺口。受干酪产能限制，发展前景广阔的乳清蛋白质在国内产量非常有限，国家应该在战略上加以重视。

参 考 文 献

Abou-Samra, R., Keersmaekers, L., Brienza, D. et al. (2011). Effect of different protein sources on satiation and short-term satiety when consumed as a starter. *Nutrition Journal* **10** (1): 139.

Amelia, I. and Barbano, D.M. (2013). Production of an 18% protein liquid micellar casein concentrate with a long refrigerated shelf life. *Journal of Dairy Science* **96** (5): 3340-3349.

Bieldt, B. (2013). Innovation, the lifeblood of the dairy industry: processing. *The Dairy Mail* **20** (12): 111-113.

Bong, D. (2013). Evaluation of an alternate processing method for Greek Style Yogurt using micellar casein concentrate: technical and business aspects. Master thesis, Cornell University.

Bong, D.D. and Moraru, C.I. (2014). Use of micellar casein concentrate for Greek-style yogurt manufacturing: effects on processing and product properties. *Journal of Dairy Science* **97** (3): 1259-1269.

Burd, N.A., Yang, Y., Moore, D.R. et al. (2012). Greater stimulation of myofibrillar protein synthesis with ingestion of whey protein isolate v. Micellar casein at rest and after resistance exercise in elderly men. *British Journal of Nutrition* **108** (6): 958-962.

Chandrapala, J., Duke, M.C., Gray, S.R. et al. (2015). Properties of acid whey as a function of pH and temperature. *Journal of Dairy Science* **98** (7): 4352-4363.

Chandrapala, J., Chen, G.Q., Kezia, K. et al. (2016). Removal of lactate from acid whey using nanofiltration. *Journal of Food Engineering* **177**: 59-64.

Chen, L., Guo, C., Guan, P. et al. (2007). Isolation of lactoferrin from acid whey by magnetic affinity separation. *Separation and Purification Technology* **56** (2): 168-174.

Chen, G.Q., Eschbach, F.I., Weeks, M. et al. (2016). Removal of lactic acid from acid whey using electrodialysis. *Separation and Purification Technology* **158**: 230-237.

Chierici, R., Sawatzki, G., Tamisari, L. et al. (1992). Supplementation of an adapted formula with bovine lactoferrin. 2. Effects on serum iron, ferritin and zinc levels. *Acta Paediatrica* **81** (6-7): 475-479.

Davidsson, L., Kastenmayer, P., Yuen, M. et al. (1994). Influence of lactoferrin on iron absorption from human milk in infants. *Pediatric Research* **35** (1): 117-124.

Dharmasena, S., Okrent, A., and Capps, O.J. (2014). Consumer demand for Greek-Style Yogurt and its implications to the dairy industry in the United States. Presented at the Agricultural and Applied Economics Association's 2014 AAEA Annual Meetings in Minnesota, USA (July 27-29, 2014).

Grant, C., Rotherham, B., Sharpe, S. et al. (2005). Randomized, double-blind comparison of growth in infants receiving goat milk formula versus cow milk infant formula. *Journal of Paediatrics and Child Health* **41** (11): 564-568.

Gurel, P. (2016). Live and active cultures: gender, ethnicity, and "Greek" yogurt in America. *Gastronomica: The Journal of Critical Food Studies* **16** (4): 66-77.

Ha, E. and Zemel, M.B. (2003). Functional properties of whey, whey components, and essential amino acids: mechanisms underlying health benefits for active people. *The Journal of Nutritional Biochemistry* **14** (5): 251-258.

Hayes, A. and Cribb, P.J. (2008). Effect of whey protein isolate on strength, body composition and muscle hypertrophy during resistance training. *Current Opinion in Clinical Nutrition & Metabolic Care* **11** (1): 40-44.

Holt, C. (1975). Casein micelle size from elastic and quasi-elastic light scattering measurements. *Biochimica et Biophysica Acta (BBA)-Protein Structure* **400**(2): 293-301.

Hurt, E., Zulewska, J., Newbold, M. et al. (2010). Micellar casein concentrate production with a 3X, 3-stage, uniform transmembrane pressure ceramic membrane process at 50 °C. *Journal of Dairy Science* **93** (12): 5588-5600.

Jones, S.B., Robbins, C., and Hansen, C.L. (1993). Sodic soil reclamation using cottage cheese (acid) whey. *Arid Land Research and Management* **7** (1): 51-61.

Kalaivani, S. and Regupathi, I. (2015). Synergistic extraction of α-Lactalbumin and β-lactoglobulin from acid whey using aqueous biphasic system: process evaluation and optimization. *Separation and Purification Technology* **146**: 301-310.

Keller, A.K. (2013). Process and system for drying acid whey. US20,150,056,358A1, filed 23 August 2013 and issued 26 February 2015.

Kuo, K.P. and Cheryan, M. (1983). Ultrafiltration of acid whey in a spiral-wound unit: effect of perating parameters on membrane fouling. *Journal of Food Science* **48** (4): 1113-1118.

Lacroix, M., Bos, C., Léonil, J. et al. (2006). Compared with casein or total milk protein, digestion of milk soluble proteins is too rapid to sustain the anabolic postprandial amino acid requirement. *The American Journal of Clinical Nutrition* **84** (5): 1070-1079.

Levenson, E. (2017). China's new two-child policy sparks increase in births. http://www.cnn.com/2017/01/23/world/china-two-child/index.html (accessed 1 January 2018).

Mawson, A. (1994). Bioconversions for whey utilization and waste abatement. *Bioresource Technology* **47** (3): 195-203.

Muller, A., Chaufer, B., Merin, U. et al. (2003). Purification of α-lactalbumin from a prepurified acid whey: ultrafiltration or precipitation. *Le Lait* **83** (6): 439-451.

Nassauer, J., Fritsch, R. Gotzmann, A. et al. (1996). Spray drying of acid whey, acid permeate and mixtures thereof. US Patent 5,580,592A, filed 30 June 1994 and issued 3 December 1996.

Nelson, B. and Barbano, D. (2005). A microfiltration process to maximize removal of serum proteins from skim milk before cheese making. *Journal of Dairy Science* **88** (5): 1891-1900.

Peng, Y., Serra, M., Horne, D.S. et al. (2009). Effect of fortification with various types of milk proteins on the rheological properties and permeability of nonfat set yogurt. *Journal of Food Science* **74** (9): C666-C672.

Roberts, A., Chierici, R., Sawatzki, G. et al. (1992). Supplementation of an adapted formula with bovine lactoferrin: 1. Effect on the infant faecal flora. *Acta Paediatrica* **81** (2): 119-124.

Savello, P., Ernstrom, C., Kalab, M. et al. (1989). Microstructure and meltability of model process cheese made with rennet and acid casein. *Journal of Dairy Science* **72** (1): 1-11.

Schingoethe, D.J. (1976). Whey utilization in animal feeding: a summary and evaluation. *Journal of Dairy Science* **59** (3): 556-570.

Semo, E., Kesselman, E., Danino, D. et al. (2007). Casein micelle as a natural nano-capsular vehicle for nutraceuticals. *Food Hydrocolloids* **21** (5-6): 936-942.

Tomita, M., Wakabayashi, H., Yamauchi, K. et al. (2002). Bovine lactoferrin and lactoferricin derived from milk: production and applications. *Biochemistry and Cell Biology* **80** (1): 109-112.

Yoshida, S. (1991). Isolation of lactoperoxidase and lactoferrins from bovine milk acid whey by carboxymethyl cation exchange chromatography. *Journal of Dairy Science* **74** (5): 1439-1444.

Zhou, S.J., Sullivan, T., Gibson, R.A. et al. (2014). Nutritional adequacy of goat milk infant formulas for term infants: a double-blind randomised controlled trial. *British Journal of Nutrition* **111** (9): 1641-1651.

Zydney, A.L. (1998). Protein separations using membrane filtration: new opportunities for whey fractionation. *International Dairy Journal* **8** (3): 243-250.

索 引

B

巴氏杀菌 22
必需氨基酸 112

C

超滤 13
超声改性 57

D

代餐食品 121
蛋白棒 119
蛋白质变性 55
蛋白质补充剂 120
蛋白质代谢 116
蛋黄卵磷脂 72
淀粉 65

E

二硫键 59

F

反渗透 13
辐射改性 58

G

干酪 3
功能特性 90
谷胱甘肽 87

果胶 66

H

化学键合理论 183
黄原胶 67

J

机械互锁理论 183
胶凝剂 127
菊粉 65
聚合乳清蛋白质 127

K

卡拉胶 68
抗氧化肽 95
抗氧化特性 90
可食用膜 100

L

酪蛋白 3
酪蛋白胶束 13
酪蛋白糖巨肽 19
酪蛋白盐 202
类胡萝卜素 135
冷冻干燥法 142
冷凝胶 59
离子交换色谱 19
磷脂酰胆碱 72
卵磷脂 71

M

酶促变性 57
酶凝酪蛋白 202
美拉德反应 62
免疫球蛋白 20
明胶 70
膜加工技术 12
膜滤 13
磨光性 182

N

纳滤 13
黏合强度 182，188
凝乳酶 4
牛血清白蛋白 20

P

喷雾干燥法 142

Q

起泡性 90
亲和色谱法 21
巯基 59

R

热变性 55
热凝胶 59
溶解性 90
乳过氧化物酶 7，20
乳化性 90
乳清 1
乳清蛋白质 1
乳清蛋白质木清漆 184
乳清分离蛋白 6

乳清粉 6
乳清浓缩蛋白 6
乳清析出 128
乳糖 6，14
乳铁蛋白 7，20
乳脂肪球膜蛋白 46

S

膳食营养素推荐供给量 120
生长因子 45
手术胶 195
酸乳清 2
酸性酪蛋白 202

T

甜乳清 2
甜乳清粉 22
脱盐乳清粉 22

W

微滤 13
微囊化 137
维生素结合蛋白 47

X

吸附理论 183
希腊酸奶 200
消化率 85
血管紧张素转换酶 36

Y

氧化稳定性 130
阴离子多糖 66
婴儿配方乳粉 99
运动饮料 99

运动营养　203

Z

增稠剂　127
支链氨基酸　84
脂肪替代品　134
纸胶水　192

致敏性　85

其他

ACE抑制肽　36
α-乳白蛋白　7
β-乳球蛋白　7